D1064762

THE CHEMICAL LABORATORY: ITS DESIGN AND OPERATION

THE CHEMICAL LABORATORY: ITS DESIGN AND OPERATION

A Practical Guide for Planners of Industrial, Medical, or Educational Facilities

by

Sigurd J. Rosenlund

np

NOYES PUBLICATIONS
Park Ridge, New Jersey, U.S.A.

Library of Congress Catalog Card Number: 86-31183
ISBN: 0-8155-1110-8
Printed in the United States

Published in the United States of America by
Noyes Publications
Mill Road, Park Ridge, New Jersey 07656

10 9 8 7 6 5 4 3 2

Library of Congress Cataloging-in-Publication Data

Rosenlund, Sigurd J.
The chemical laboratory.

Includes index.
1. Chemical laboratories--Design and construction.
2. Chemical laboratories--Management. I. Title.
[DNLM] : 1. Chemical industry. 2. Facility Design and
Construction. 3. Laboratories--organization &
administration. QD 51 R814c]
QD51.R57 1987 542'.1 86-31183
ISBN 0-8155-1110-8

Preface

It is my hope that this book will fill a gap on the technical library shelf by offering help to those involved with either planning new laboratories or expanding existing ones. It is based on over thirty years of laboratory experience, including day to day operation, design of new facilities, supervision of construction, and consultation. I have witnessed the fruits of good planning and the unfortunate consequences where planning was inadequate.

Who can profit from such a book? The supervising chemist who must define the needs of a new laboratory will find many practical suggestions. So will the administrator looking for ways to justify a facility that will not become outdated in a few years. The designer or engineer will be better able to see things from a client's viewpoint, as will the contractor in charge of certain aspects of construction. The doctor or dentist setting up a laboratory facility next to the office will discover suggestions for making the best use of limited space. The supplier of laboratory furniture and equipment will find new ways to advise his customers. Finally, the young chemist who finds himself charged with starting up a new operation will have the reference I wished for when I was in that position many years ago.

Above all, this book is intended to be a *practical* guide to laboratory planning. It will not go into the more sophisticated areas of science and technology, Instead, it will deal with a broad variety of more common matters, some of which may be overlooked or underestimated by the laboratory planner.

Perhaps an explanation should be offered here for my use of "he" throughout the book in referring to the person in charge of daily laboratory operation. This pronoun is used in its traditional sense to refer to either a man or a woman. The newer "he or she," while more accurate, considering the many women in charge of laboratories today, is also more cumbersome and has been avoided for that reason.

I would like to thank all those who have given their time and thoughtful comments. Benjamin F. Naylor, chemistry professor emeritus of San Jose State University, read the manuscript in its early stages and contributed valuable information on educational laboratories. Alan C. Nixon, past president of the American Chemical Society, along with other members of Calsec Consultants, Inc., offered helpful suggestions from their diversified experiences. Numerous laboratory personnel guided me through their facilities, and distributors of laboratory products kept me informed about their latest products. Plumbers, electricians, and others in the building trades had many practical hints.

Finally, I would like to thank my wife, Barbara, for her encouragement, suggestions, and countless hours of editing.

January 1987 Sigurd J. Rosenlund

NOTICE

Contents

Introduction

A laboratory may come in any size or shape. It may be a room in an industrial plant, a wing of a hospital, or a whole building on a college campus. All of these present similar problems and decisions at the planning stage. Where should the laboratory be located? How much space is required? Will a proposed layout contribute to smooth traffic flow? What utilities are needed? What safety factors should be built in? These are just some of the major questions planners must address.

The results of poor planning usually do not show up until a facility has been in operation for some time. By then, correcting them is invariably expensive. Anyone who has worked in a laboratory for even a short period has seen some of these problems. There may be overcrowded workbenches, where permanent equipment and set-ups leave little or no room for non-routine work, or inadequate wiring which requires the use of cumbersome and hazardous extension cords on a permanent basis. Poor ventilation is a common problem, causing both discomfort and hazards. Improperly chosen bench tops may be stained by chemicals or damaged by heat. Many laboratories have awkward traffic patterns, resulting in wasted time during performance of routine tasks. Storage areas may be inadequate or poorly located. The list goes on and on.

Many will blame such problems on lack of funds when the laboratory was built. This may not be the case. A well-planned and efficient laboratory does not have to cost more than a poorly planned one. It is mostly a matter of putting the available money to work where it counts most. This book gives many examples where money can be saved without causing operational problems later. It also presents cases where additional money spent at the outset has paid off in a safer and more efficient operation for years to come.

Who plans and builds a laboratory? In a small facility the whole job is often handled by in-house talent. I have seen many cases where such talent was capable of taking on the challenge. I have also seen numerous cases where professional assistance should have been employed. This book will help the do-it-yourselfer decide when such assistance is needed. At the other end of the spectrum, a design or engineering firm may be hired to do the job on a turn-key basis. Such a firm can guarantee professional results, but will these be specifically tailored to the needs of this particular laboratory? Examples of both underdesign and overdesign, usually resulting from poor communication between designer and client, are given.

Throughout the book, the person in charge of day-to-day operations is referred to as the *laboratory operator*. This is not an administrator or supervisor located in an office down the hall or in another building. The laboratory operator must be heavily involved in all aspects of planning. Only he can estimate space requirements, check a proposed layout for practical and safe operation, and recommend allowances for future expansion. Regardless of the amount of professional assistance available, the laboratory operator can expect to burn much midnight oil. During construction he must be available at all times to take care of those numerous problems nobody had predicted.

This book not only deals with major matters, such as laboratory size and location, layout, and utilities. It also includes seemingly minor topics, such as choice of paints and floor coverings, money saving hints for utility hookups, and types of ceiling treatment.

Safety and waste disposal are treated in detail because of their ever increasing importance.

Planning and building a laboratory requires a cooperative effort involving administrators, designers, equipment supply houses, contractors, and the laboratory operator. A laboratory designed for efficient operation can be achieved only if all of them work together with mutual respect and the best possible communication.

1

Preliminary Planning

Once it has been decided that a new laboratory should be built, some important basic planning must be done. Whether or not an architect or designer is to be called in later, those in charge of the laboratory operation will need to consider questions such as these:

> What types or work will be performed both initially and in the foreseeable future?
>
> Will any of this work create special hazards?
>
> What equipment will be required?
>
> Will any of the work produce excessive fumes, heat, dust, or noise?
>
> Should any of the jobs be performed in isolated areas or in separate rooms?
>
> Will any tasks require a controlled environment?
>
> How much room will be taken up by permanently installed equipment?
>
> How much space will be needed for undesignated work areas?
>
> Where should the laboratory be located relative to other facilities?

Are there any special security requirements?

Facing such questions head-on from the earliest planning stages and making notes as information is gathered will help to avoid unpleasant surprises later on.

LISTING OPERATIONS

It is important to make a list of every task that will be performed in the laboratory, down to the smallest detail. Operations such as pH measurements, transfer of flammables from safety storage to shelf bottles, or recording observations must not be overlooked. Even in a small laboratory, the number of individual tasks will be quite substantial.

Each operation on the list should then be evaluated for problems it might create and for any special requirements. These might include the following:

Hazards the operation may create and what precautions must be taken (fume hood, separate room, etc.)

Non-hazardous nuisance it may cause (odor, dust, heat, noise, steam, etc.)

Possible contamination of other work being performed.

Vibration or other disturbance of other operations.

Special environment needed (controlled temperature, clean-room conditions, absence of drafts, etc.)

Security requirements (controlled access to certain instruments or operations, etc.)

It will take time to come up with all of this information, particularly in cases where new types of work are contemplated. If a new piece of equipment is to be installed, planners should obtain as much literature as possible from the manufacturer and make a careful check of procedures for which it will be used. If it can be arranged, a visit to another laboratory in which this instrument is already in use will prove very helpful.

It is not only the new work that will need evaluation, however. Even operations that have been performed for many years should be reviewed and updated as needed. Safety requirements, for example, could have been changed, as will be discussed in Chapter 4.

A partial check list of operations for an industrial chemical laboratory is shown in Table 1. The format of a formal list will vary considerably from one laboratory to another, but with such an aid, one can easily see which operations are compatible and then group these together. Those that need special treatment will readily stand out. The planner will also be able to estimate the number of rooms required for the total operation. Finally, a complete list of all laboratory functions will facilitate the next step, an estimate of space requirements.

ESTIMATING SPACE REQUIREMENTS

A typical laboratory that has been in operation for some time usually has run out of space for optimum operation. In some cases, the space may be there but cannot be utilized to full advantage. Work benches gradually get covered with permanent equipment set-ups, leaving little room for other work. Lack of storage space for supplies and samples becomes the rule rather than the exception. Adequate room for a desk, bookcase, or typewriter has often been overlooked. As more personnel is added, these problems become critical. Overcrowding also has a serious effect on safety.

Since such conditions are evident so soon in many cases, it is obvious that they could have been avoided by more careful planning. The most important space requirements to consider are

Table 1: Checklist of Operations

Operation	Hazards, Problems								Conditions Required								
	Fire	Corrosive fumes	Toxicity	Smoke, odor	Heat	Vibration	Noise	Dust	Separation	Fume hood	Special fume hood	No corrosive fumes	No vibration	Special ventilation	Closeness to sink	Explosion proof environment	No dust
Analytical weighings									X			X	X				X
Ether extractions	X								X					X		X	
Powder screening						X	X	X	X								
Muffle furnace				X	X									X			
Drying oven				X	X									X			
Solvent evaporation	X			X	X					X							X
Solvent dispensing	X								X					X		X	
Solution preparation		X	X							X					X		X
Perchloric acid digestions		X	X								X						
Kjeldahl digestions		X	X	X	X				X					X	X		
Desk work, telephone									X				X				X
Vacuum pump						X	X		X								
Microscopic work									X			X	X				X
etc.																	

those for work benches, free-standing equipment, office space, and storage. In addition, due attention must be paid to the fact that some work areas must be separated to avoid hazards or contamination.

Work Bench Space

The first step in space planning should be to estimate how much bench space will be required. Benches must accommodate various pieces of permanently installed equipment and still have room for both frequently performed and special one-time tasks. Benches may be either installed against the wall or placed back to back in peninsulas or islands. The exact configuration will be worked out later. For now, what is important is the total number of running feet of bench space that will be required.

All bench-mounted equipment should be listed with its dimensions, which can be taken from measurements or from catalog data. It should be noted which instruments may have to be put at a certain minimum distance from other objects in order to avoid interference or allow for servicing. Work space should be allowed next to each instrument for samples, notebooks, etc. This space may be considerable in case of an analytical instrument on which many samples are to be tested at one time. Space sharing should be discouraged, with the required work space next to an instrument reserved for that alone. Sinks and fume hoods should also be included in this list, with an allowance of at least 18 inches of free space on each side of a sink.

When all equipment has been listed, the total number of running feet needed can be added up. How much more will be required? This will vary from one laboratory to another. In an industrial laboratory where both routine testing and product development work are to be performed, this figure can safely be multiplied by four for a realistic estimate. In some laboratories used exclusively for specific types of work with no other types contemplated, this figure may be lower. It should be kept in mind, however, that expanding an existing laboratory is an extremely expensive undertaking. It is desirable, therefore, to have some unused wall space and some free floor area when a new laboratory begins operation.

Free-standing Equipment

There is more to a laboratory than work benches and the instruments mounted on them. Free-standing equipment must also be considered. This includes refrigerators, safety storage cabinets for chemicals, safety shower, desk space, typewriter stand or computer terminal, or any other equipment that is not bench-mounted. File cabinets, which are real space-robbers, must not be forgotten. In one laboratory, much space was saved by placing two-drawer file cabinets beneath the large table used for sorting samples.

Arguments for Additional Space

Careful scrutiny of all these measurements will soon show a laboratory planner that much more space is required than was originally thought. Next comes the competition for this space, at which time convincing evidence must be produced. One laboratory manager proudly showed off his new facility, the product of such a battle. Yes, he had been accused of taking up far more space than he really needed. He had done his homework, however, and presented management with enough data to convince them. Although his laboratory had much free floor space and a whole empty wall, he had looked into future plans and knew what would be required before too long. He also had spare circuits in the breaker box and plumbing which could readily be expanded.

EDUCATIONAL LABORATORY REQUIREMENTS

Figuring space for educational laboratories requires some different considerations. For a given course, a certain amount of bench space must be allotted to each student. Only those in charge of the course are able to make a realistic estimate of this. In addition, since each work station is used by more than one student, there should be sufficient space beneath it or close by for individual lockers or drawers. Organic laboratories usually require more space per student because of the equipment set-ups used and the need for separation to avoid fire hazard. The total amount of bench space thus depends on the number of students each room is designed for.

Fume Hoods

In an educational laboratory, fume hood space requirements are substantial, since so many students need a hood at the same time. Many operations that used to be performed on the bench must now be done in a hood for safety reasons. Inadequate hood space has caused many problems even in fairly modern university chemistry buildings.

Analytical Balances

Most educational laboratories keep their analytical balances in separate rooms, which are locked when not in use. Balances less than three feet apart are crowded, so sufficient space must be allowed. Such rooms will also be useful for other instruments, such as spectrophotometers, that do not give off heat or fumes.

Other Equipment

Large capacity balances, centrifuges, and similar equipment are generally placed on a separate counter away from the benches. The size of this counter must be carefully estimated. There must also be room for equipment and supplies kept in the laboratory as opposed to the stockroom. Of course, space must also be allowed for a safety shower and eyewash station.

Planning for the Future

There was general rejoicing in a community college chemistry department a few years ago when funds were granted for a long desired expensive analytical instrument. Then someone realized there was really no good place to put it, so a make-shift arrangement had to be made. Planners of this nearly new facility should have considered such equipment additions in their preliminary calculations, in anticipation of the day when funds might become available.

STORAGE AREAS

When it comes to laboratory storage, it is safe to say that the space required is at least twice what a planner would estimate. Samples, reagents, and spare equipment will pile up at an alarming rate. To this should be added the fact that some items will have to be stored under controlled conditions and that valuable items will need to be kept in locked storage. Flammables, even in moderate amounts, need special storage. Since regulations vary from one area to another, this matter should be discussed with local fire department officials.

Industrial, Medical, and Research Laboratories

A manufacturer usually stores samples of both raw materials and finished products for extended periods of time. He may even be required to do so by law. Sample storage may present hazards often overlooked. A single bottle of perfume, for instance, is too small to be considered a fire hazard, even though the material is quite flammable. Hundreds of these bottles, however, stored away as retain samples by a cosmetics manufacturer will become a fire hazard.

Medical and research laboratories have much the same storage requirements as industrial laboratories. All need adequate space for chemicals and supplies, as well as for equipment that may be used only occasionally for special purposes. They also need space for a rapidly growing number of samples.

Most storage will require shelving. Unless large items are to be stored, such as instruments not regularly in use, shelves should be no more than 12 inches deep. A common distance between shelves is 12 to 15 inches, although this may be less for some sample storage. Such figures will tell a planner how many feet of running shelf space is needed. A realistic estimate is an important part of preliminary planning.

Educational Laboratories

Very few educational laboratories that have been in operation for awhile have adequate stockroom facilities. Every stockroom supervisor has tales about lack of space.

Reagent Storage. Supplies of reagents are often purchased in large quantities to last for a semester or an entire year. These normally need the protected storage of the stockroom. There must be adequate space for flammables and for materials that could give off hazardous, corrosive, or unpleasant fumes. Bottles of hydrochloric or nitric acid often cause corrosion even if reasonably well closed. For some courses, certain reagents are sent out to the student laboratories for experiments and then taken back to the stockroom. Leaving them out on a permanent basis could be both impractical and hazardous. All these materials should be listed and space requirements estimated.

Glassware. Unlike reagents, glassware can be stored almost anywhere there is room. Rather than take up valuable laboratory or stockroom space, it may be possible to store full cartons of glassware in a warehouse or another building and to bring only enough to the stockroom to take care of current needs.

Instruments. Some instruments are checked out to students only a few times each semester. These may include pH meters, small spectrophotometers, and other items. They are in storage the rest of the time. Platinum electrodes and other valuables usually need locked storage.

Preparation Space. The stockroom should have space for preparation of solutions and other items, such as unknowns for courses in qualitative analysis. This requires a regular work bench with sink. There must also be room for prepared solutions to be dispensed to student laboratories in bench-sized bottles, which take up a good deal of space.

Repair and Maintenance. Another job for the stockroom personnel is to do minor repairs and routine maintenance of equipment. Adequate bench space should be set aside for this.

Equipment Check-out. A large table next to the check-out window can be very valuable. Prior to class, equipment needed can be taken from the shelves and lined up on the table, ready for quick delivery to the students. When returned, it is placed on the table and taken back to storage after the end of the period rush is over.

The above considerations do not give a direct answer to how many square feet of space a given laboratory will need. They merely show what has to be accommodated. At this point, a planner may be able to see major discrepancies between allotted space and required space, which must be resolved before going on.

LABORATORY LOCATION

The exact location of a laboratory within a building or a complex of buildings is often the result of a grand compromise. Sometimes the laboratory planner is presented with a location and must simply make the best of it. If the location is poor, he will then have to use his best persuasive powers to bring about a change. What arguments are effective against a poor location? If insufficient space is the problem, a careful estimate of space requirements, prepared as described, will help the planner prove his point. In other cases, the cost of conversion can be presented as a forceful deterrent. A medical laboratory, for example, is often located in a multi-purpose suite of a medical center. Such a suite may not be adaptable for laboratory use without very expensive modification.

Safety Considerations

In disputes over location, safety makes an impressive argument. For instance, there must be a safe exit through parts of the building where a fire is not likely to develop. Unfortunately, this rule is not always followed in industrial laboratories. Consider the industrial laboratory which had only one door and no windows. As if this were not enough of a safety violation, the door was directly across a narrow hall from the boiler room door. A boiler accident in this building could have entombed the laboratory staff. It is hard to believe this layout was designed by an architect and approved by local building authorities in the 1950's. While such a design would be unlikely to be approved today, small industrial laboratories are still sometimes installed without a proper permit. Such installations often violate safety rules with respect to both location and layout.

Efficiency Needs

Ease of communication between the laboratory and other areas is important to consider. There should not be a long hike to the office, processing area, or other parts of the building with which the laboratory has frequent contact. In one newly built plant, samples had to be carried down two flights of stairs via heavy fire doors at each end, across a busy production area, through another fire door, and finally to the laboratory. Impractical? Of course. Yet this layout was made by a reputable engineering company with long experience in the industrial field. Better communication between the designer and a knowledgeable company representative could have avoided this inefficient plan.

In another plant, the industrial laboratory was installed close to the processing area down a short flight of stairs. Before this location was chosen, possible hazards and environmental effects were studied. Here it took little over one minute to bring samples to the laboratory or for laboratory personnel to be on hand to investigate manufacturing problems.

Environmental Considerations

Environmental effects are often underestimated during preliminary planning. Dust or fumes entering the laboratory each time the door is opened, for example, will certainly create trouble, as will high temperatures in the area adjacent to the laboratory.

Vibration. A less obvious problem than dust, fumes, or heat is vibration, which may cause difficulties with some types of laboratory equipment, such as analytical balances. Vibration can also interfere with microscopic work, particularly if this is combined with photography. In industrial plants, operation of heavy equipment may cause considerable vibration and should be considered when laboratory location is determined.

One research laboratory was located on the second floor of a building in which a diaper laundry occupied the first floor. The laundry equipment would periodically send veritable shock waves through the building, making many laboratory operations impossible for a short while. The laboratory workers referred to these annoying incidents as the times when the laundry "dropped its load."

Possible vibration sources outside the building, such as nearby railroad tracks, should be considered also. In a new testing laboratory, heavy truck traffic immediately outside caused periodic vibration problems, even though the building sat on a solid concrete slab. Had this laboratory been on an upper floor, the vibration would have been even more severe.

Sunlight. There is always a question of whether or not a laboratory should have windows. Windows take up valuable wall space, always at a premium, and should not be counted on for providing effective ventilation. While sunlight may be a good morale booster for workers and assist in keeping the building warm on cold winter days, it is at best a nuisance when it shines on the work benches. If there are to be windows, they should preferable face north or east.

Noise. Noise is another environmental factor to be considered. It does not have to be very loud to seriously affect worker performance if it is persistent. The source of noise disturbance may be one of many things—plant equipment, heavy traffic outside, ventilation fans, etc.

Access to Utilities

The ease of access to utilities should also help determine the laboratory's location. Most laboratories will need hot and cold water, electric power, gas, and sewer connection. The cost of bringing these to a remote location may be high. The sewer is often the most problematic. Many a concrete slab has been torn up at considerable expense in order to install a laboratory sewer. A building professional should be consulted to give advice in such cases.

Zoning Regulations

Finally, is it permissible to build a laboratory in the proposed location? Some regulation could make the project impractical or even impossible. Generally there is little problem getting permission on a college campus or in an industrial plant, although this must be checked out with the proper authorities. The situation becomes quite different if a new building is to be erected for laboratory use or if an existing building is to be converted. Zoning regulations must be considered and an opinion from local planning authorities must be obtained before further planning can take place.

2

Laboratory Layout

Time has now come to convert the preceding data and ideas to a workable laboratory layout. In all but the simplest cases this should be done with the assistance of a contractor, designer, engineer, or other building professional. Many small laboratories are planned strictly on a do-it-yourself basis in order to save money. Unless an experienced in-house person happens to be available, the results are often poor and the savings questionable. In other cases, the whole job is turned over to an architectural or engineering firm on a turn-key basis with little involvement by the laboratory operator. This sometimes results in a facility that is not altogether suitable for its intended functions, in spite of a highly professional design. Some designers tend to rely heavily on standard layouts which they have used successfully in the past but which may not be suitable for a given installation.

In other words, the laboratory operator is still the key person at this stage and should be prepared to burn a considerable amount of midnight oil, even with the best professional assistance. He will be in constant contact with the architect or designer, and the project will sometimes be like a ball that is tossed back and forth. In this game, the building professional will have the knowledge of practical and economical construction methods, as well as of local building codes. The laboratory operator will be familiar with the work to be performed, often hard to explain in detail to the designer.

LIMITATIONS

Laying out the laboratory will be accomplished under one of three conditions. In the first, and most desirable, the laboratory will be incorporated into a building still in the planning stage. In this situation, there will be some leeway in organizing the shape of the area, even though the location and overall space allowance may already have been determined. Somewhat less desirable, but quite workable, is placing the laboratory in a partitioned off section of an existing building. In either of these cases, the laboratory operator, in close cooperation with the designer or architect, will have to spend much time selecting the best room sizes and dimensions. The greatest challenge comes in the third situation, in which one or more existing rooms are simply designated by management as "laboratory" and must be utilized to their best advantage.

Whatever building limitations the planner finds himself working under, he must make certain that there is indeed enough square footage available for the laboratory. If not, this is the time to call for major changes. If the space seems large enough based on preliminary estimates, planning for the best utilization of this space may now proceed. Detail drawings will show if not only the size but the shape of the room will suit the requirements of a laboratory.

MAKING THE SCALE DRAWING

The first step in establishing the layout is to obtain or prepare an exact scale drawing of the laboratory area. For a building in the planning stages, the architect's drawing can be used. For an existing building, any drawing already available should be checked to see whether any changes have been made in the building since the drawing was made. If no drawing exists, careful measurements must be taken.

A suitably sized drafting board is a good investment, as are some basic drafting supplies. If a copy service for large size paper is available at a reasonable price, the drawing should be done on regular drafting paper, which may be purchased in either tablet or roll form. Otherwise, the drawing may be made on several sheets of regular size paper that can be reproduced on an office copier. With accurate registration marks, joining the sheets after copying is not difficult.

A typical scale for such drawings is ¼ inch to the foot, but any convenient scale may be used. On the drawing, all existing details—doors, windows, wall protrusions (common in the popular tilt-up construction), pipes and conduit on the walls, sewer outlets, utility connections, etc.—should be noted. It is important that all details be accurately recorded.

One of the copies should be designated as "master." No item should be added to this before being finalized. On the other copies, adding the desired features will be a matter of trial and error on paper, with many copies ending up in the waste paper basket.

ROOM ORGANIZATION

Sketching on copies of the drawing may now start. If walls are to be put up, a thickness of 6 inches may be assumed, unless special circumstances are present. The preferable door width is 36 inches, but 30 inches may be acceptable under some local codes. Installation of large equipment, however, could require larger (double) doors, generally five feet wide. For safety reasons, a laboratory must have two exits. When a new laboratory building at the University of California at Berkeley was in the planning stage, the building manager discovered that under new regulations, doors could no longer swing out into a hallway. They would have to open into alcoves so that the hall would be unobstructed with all doors open. With a number of rooms planned for the building, this added up to a significant floor area that could not be used for other purposes. From the standpoint of safety, however, it was of great importance.

Work Patterns

Using the space requirements previously developed, tentative placement of specific pieces of equipment may now be made. Planners should try to visualize the daily laboratory routine while doing this. Anything that will save steps and minimize congestion should be incorporated. The list of operations described in chapter 1 will make it possible to group together compatible tasks for highest efficiency and greatest safety. If planners think in terms of the distance laboratory workers will have to walk when performing routine duties, a pattern of interconnected work areas will soon begin to emerge. After some trial and error, a reasonable pattern will be developed. This should then be presented to the architect or designer, who may have some critical comments about lack of practicality from a building standpoint or conformance to codes. This will mean back to the drawing board. After several trips back and forth, a workable plan can be developed and agreed upon.

WORK BENCH DIMENSIONS

Typical laboratory benches are 36 inches high and 30 inches deep when mounted against a wall. This includes the average 7 inches (more or less than this may be required) behind them for carrying utilities. The actual base cabinet depth is 23 inches, which makes narrower counters possible in areas where space for utilities can be sacrificed. If benches are formed into a peninsula or an island, a typical total width is 54 inches. It may be more if the center part has shelves or extra utilities.

The distance between work benches should not be less than four feet. In one laboratory an extra bench was fitted into a large room by decreasing this distance to three feet, but the result was serious congestion. In educational laboratories, five feet would be advisable because of heavier traffic.

BENCH CONFIGURATION

The shape of the room will determine the location of the work benches. In a long narrow room, they may be conveniently placed along the wall, as in a Pullman kitchen. In a wider room, islands are practical but present special problems. The utility hookups, for example, may be difficult, particularly in an existing building, unless there is easy access from below. If a sewer connection is required, it must be made below the floor level. Incoming utilities can be brought in from above the island through a chimney-type arrangement going to the ceiling.

Peninsulas may be a better solution than islands in rooms that can accommodate them. In such an arrangement, utilities are run along the wall and branched off as required. However, if a regular work bench is installed along this wall also, such a plan will create large, less useful corner areas. One laboratory solved this problem by substituting a 12-inch wide work surface along the wall in place of a bench. Plumbing and wiring were installed below, and the peninsulas branched out from there. The exposed utility space was then covered with removable plywood panels painted to match the furniture. This eliminated wasteful corners, while the narrow work surface proved useful for many jobs.

Educational laboratories often have island work benches, generally with a sink at one or both ends. Depending on room size, a peninsula arrangement could save considerable cost with no loss in efficiency.

Although they are not too practical as work surfaces, corners can be put to good use accommodating large items, such as drying ovens. There are also fume hoods made to fit into corners. Since a corner is easily reached from the work areas on both sides, it can also be a good place for a sink.

STORAGE CABINETS AND SHELVES

Laboratory storage cabinets and shelves are available in different widths and are usually 12 inches deep. A depth of more than this is not recommended unless large, bulky items are to be stored. Narrower shelves, six to nine inches deep, have been found more practical for reagents and other small items. If shelves and cabinets are to be mounted on walls above work benches, possible interference with work performed there must be considered. Three feet or more should be allowed for aisles.

HEAT-PRODUCING EQUIPMENT

Some types of equipment give off a considerable amount of heat. Most planners are aware of this and will provide for appropriate ventilation, a matter that will be discussed in detail later. Radiant heat, however, is less often recognized as a problem. In a food laboratory, for example, a six-unit Kjeldahl digestion and distillation apparatus was installed against a wall and the hot air was drawn off overhead. The heat radiating from 12 flasks and heaters, however, made the workers on the other side of the narrow room very uncomfortable. Another laboratory solved this problem by installing 12 separate digestion and distillation Kjeldahl units along the side walls of an alcove, where they radiated against each other rather than into the room. Even though it was quite hot for a worker standing between them, the time spent there was limited and other operations were not affected.

Muffle furnaces also produce radiant heat but only during the brief periods when they are open.

TOXIC AND FLAMMABLE MATERIALS

Areas for handling toxic or flammables should be segregated. A chemical like acetone, for example, should never be handled in the vicinity of an open heat source. Work with highly toxic materials should be strictly confined to designated areas. Many operations must be performed in fume hoods. For others, only improved ventilation may be required.

ANALYTICAL BALANCES

Analytical balances are among the prima donnas of the laboratory, requiring separate and unequal treatment. They refuse to cooperate if there is the least amount of vibration and will quickly expire upon exposure to corrosive fumes. Yet they must at all times be close to the action, or laboratory workers will have to do a fair amount of hiking. As a rule, analytical balances are placed on separate tables, which should be large enough to also hold a desiccator for samples and the operator's notebook. The table must be as stable as the rock of Gibralter. In a teaching laboratory, balances are usually placed in a separate weighing room, which can be locked when not in use. This can be located between two laboratories, giving it better accessability.

SAMPLE RECEIVING

Analytical laboratories need an area where incoming samples can be sorted and recorded. The size of this is hard to overestimate. In addition, some laboratories need an area where samples can be prepared for analysis. A pesticide laboratory, for instance, may want to set aside a complete room for such work, since it is often quite messy.

OFFICE SPACE

Two types of office space are usually needed. One is the area where laboratory workers perform calculations, check procedures, and make up reports. The other is for the laboratory supervisor to do desk work and hold conferences or telephone conversations, which may require privacy or quiet.

Workers' Area

The laboratory workers' office space should be as close as possible to work areas and to frequently used files. There should also be room for a typewriter, if required. The work may be done on a separate desk or a section of desk-height work bench.

Supervisor's Area

A supervisor's office may be a separate room, although an area with partitions extending part-way to the ceiling is often just as satisfactory and less expensive. In addition, such partitioning offers the bonus of valuable extra wall space. A large window between office and laboratory is recommended for good supervision of activities. Since books and reference materials are often kept in this office, adequate space for shelves must be provided. In a small laboratory without much interference, the supervisor may simply need a desk in a corner of the room.

Furniture Dimensions

A typical office desk is about 3x5 feet, though another size may be more desirable. There should be at least three feet between the desk and the wall for getting in and out of a chair. Typical dimensions for file cabinets are 15x25 inches, but the opened drawers may increase the total depth to as much as 48 inches. Bookcases are usually 9 to 12 inches deep and are available in many widths.

SAFETY SHOWER

The placement of a safety shower and eyewash station must be given careful thought. It should be directly accessible and no more than ten seconds away from any work area. In some cases, it may be placed just outside the laboratory in a spot where it can also serve other parts of the building. Educational laboratories or other large, multi-room facilities may need a shower in each room.

COMPLETING THE LAYOUT

Making the completed layout is much like solving a puzzle where pieces have to fit together in a certain way. In fact, some planners like to cut out pieces of heavy paper representing the various laboratory components and juggle them around until a reasonable layout is obtained.

Gradually all things that are to go into the laboratory, such as work benches, refrigerators, safety storage cabinets, floor-mounted equipment, desks, file cabinets, balance tables, and a myriad of other items, are in place. Now is another good time to check on traffic patterns. Will a piece of equipment stick out too far? Will a desk chair cause obstruction? Will the refrigerator door cause problems when open? These are just a few of the questions that should be asked again.

Now the exact locations of sinks and places where gas, DI water, or other utilities will be needed should be marked on the current drawing, and the location of instruments using electric power should be shown.

Still another check on safety should be made. Are exits readily accessible? Is the safety shower easy to reach? Are areas for handling hazardous materials properly segregated? Are areas of potential hazards away from important traffic lanes?

When all these questions are answered to the planner's satisfaction, all details can be transferred to the master copy of the drawing. Even though the architect or designer will not be directly concerned with the equipment to be moved in, this should be shown on the drawing. One way to show this is with dotted lines. This will also make it easier for the lighting designer, as it will show up the areas where good light is required and where shadows will be a problem.

The designer will no doubt find flaws here and there. Some objections may be caused by a lack of familiarity with the work to be performed, a problem which can be solved by good—and frequent—communication.

Based on the material submitted on the master copy of the drawing, the designer will prepare working drawings to be used by the contractors during construction. The laboratory planner should not be afraid to ask questions about anything on these drawings that does not seem absolutely clear. Misunderstandings are common, and this is the point to clear them up.

3

Utility Requirements

A typical laboratory will require a variety of common utilities, such as water, gas, and electric power. There must also be sewer connections and adequate provision for ventilation, as well as means for heating and cooling the area. In addition, there are often special laboratory needs, such as compressed air and vacuum.

At this point of the planning, we are mainly concerned with quantitative estimates of such utilities and the ease, or lack thereof, with which they can be brought to the laboratory area. These considerations have a profound effect on laboratory layout. In laboratories that have been in operation for some time, a lack of needed utilities is a major problem, particularly when new procedures are introduced. Estimates should be made by the laboratory operator in close cooperation with building professionals and should take future needs into consideration.

HOT AND COLD WATER

Most laboratories have relatively modest requirements for water, but it must be available in the right places. Some planners find it useful to terminate rough plumbing near a corner of a room, thereby making it easier to extend the finished plumbing along two walls. Rough plumbing may be run below the floor, inside the walls, or in the ceiling. Plans for this should be made in conjunction with the detailed room layout, as described in the preceding chapter. Determining the number of sinks needed and their approximate location at this point, moreover, will be of great help to the designer.

The laboratory operator should be able to assist a professional estimator determine the water heater size from his knowledge of work to be performed. It is often practical and economical to have one heater serve a fairly large area, as long as the distance from heater to points of usage is not excessive. A testing laboratory which was located in a long, narrow building, on the other hand, found it more suitable to install two smaller heaters, one at each end of the building. Insulating hot water lines is highly recommended.

In hard water areas, a water softener will increase the life of water heaters and plumbing fixtures. It must be kept in mind, however, that a softener does not decrease the total mineral content of water; it merely modifies it.

DEIONIZED WATER

A still used to be a fixture in laboratories. Older chemists have unpleasant memories of the occasional cleanouts that needed to be made. Today laboratories have found that deionized water will serve most of their purposes just as well at a much lower cost. Rental-type deionizers are available from several companies, who will install the resin tanks and exchange them as needed. There is usually an installation charge, a fixed monthly charge, and a separate charge each time a tank is changed. The service company will make an estimate as to what size tank or tanks a laboratory needs.

The use of two tanks in series is highly recommended. When one gives out, the other will take over until the service company arrives. This avoids annoying delays that are bound to occur should a single tank stop operating at some crucial moment.

The condition of a DI water tank is indicated by a small neon light. When the light goes out, the tank is exhausted. Since the power consumption of the light is very low, it can be plugged in to any available circuit.

DI water is not equivalent to distilled water in all ways. While the content of most ionic species is very low, it does contain dissolved gases, such as air and carbon dioxide. The latter caused problems in one laboratory when DI water was used for diluting poorly buffered samples for pH measurements. Erratic results were also reported in another case when DI water was used in connection with the determination of trace amounts of boron. The manufacturer explained that when close to exhaustion, the resin used would no longer be effective in holding back traces of this element.

Finally, DI water is anything but sterile. In fact, the resin beds seem to support bacterial growth quite well. In other words, it does not replace distilled water for all applications, but those who use it right will enjoy pure water at reasonable prices.

DI water is handled in plastic pipe, generally PVC. For rough plumbing, it is usually desirable to have the pipe terminate close to the hot and cold water lines. Tanks should be located where servicing will be convenient and where minor water spills during tank exchange will cause no problems. One laboratory placed them in a small closet with a door leading to the outside of the building, where the service truck could park directly in front of the door.

DISTILLED WATER

There are times when DI water just will not do, as seen in the above examples. Distilled water is then required. A laboratory may distill some of its DI water, which will give the still a longer life. On a large college campus, steam was available from the central heat generating plant, an ideal source of distilled water. Then there are cases where ultra-pure water is required for special purposes. This is best produced from previously purified water in all-glass apparatus under carefully controlled conditions.

GAS

For many years, gas was the primary source of heat in the laboratory. Today, although electric heaters have become more common, gas still has some advantages sometimes overlooked. It is a quick source of heat and very fast to regulate. A gas burner can be pulled away in a fraction of a second should a distillation or a reflux operation get out of hand. In case of a spill, a gas burner is quickly dismantled and cleaned, whereas an electric heating element may have to be replaced. Another advantage is the lower cost of the gas burners.

Running gas lines into a laboratory is not difficult once it has been determined where they are needed. It may simplify plumbing installation to have rough plumbing for gas terminate close to the water lines. When estimating total requirements for gas, it should be kept in mind that it may also be needed for a water heater and a heating unit for the building.

ELECTRIC POWER

In laboratories in operation for some time, there are often not enough electrical outlets or enough circuits to handle the load. Retrofitting for more power is very expensive. One laboratory received several bids for wiring their new facility. The accepted one was much lower than the others because money was saved by installing the bare minimum number of circuits and no spare circuit breakers. Every piece of conduit was jammed full of wires. Within two years, more wiring was needed for new equipment. This very costly expansion would not have been necessary if allowance had been made for future work.

From the list of laboratory operations previously prepared, it will now be easy to single out the ones that require power. The manuals for the equipment already on hand and catalog information on items yet to be purchased can provide the power requirements. Compared to what was available some years ago, modern laboratory equipment does not need much power. Exceptions are heating devices and motors, which may be very power hungry. A list of the wattages involved should be made, noting which equipment operates on 110 volts and which on 220. Allowance should be made for future purchases of equipment. This information will help the electrical engineer or contractor determine the number of circuits.

Most modern instruments will operate over a wide voltage range but they may, at the same time, be quite sensitive to voltage variations during use. Such variations often throw calibrations off, as they did in one laboratory where an AA spectrophotometer suddenly began to act up. Investigation showed that someone had plugged a large hot plate into the same circuit in another room. As the proportional switch in the plate went on and off, the circuit suffered enough voltage variation to affect the instrument. High wattage equipment was from then on kept off this circuit.

Knowing about this case, another laboratory planner insisted on a special circuit to be used only for instruments drawing low and even power. Still another laboratory managed to justify a very elaborate voltage stabilizing device to feed instruments, which they felt had paid off.

SEWER CONNECTION

In a building under construction, connecting a laboratory to a sewer is usually not too much of a problem, but it can be a costly affair in an existing building. This is especially true in a one-story facility with a concrete floor. Communities with strict waste water regulations may require a sampling station for laboratory waste prior to the point where it joins the building's sanitary waste. In other cases, they may demand a large underground mixing tank before the point of entrance to the public sewer. The laboratory operator would do well to personally check with local authorities on this matter and discuss it with the designer.

COMPRESSED AIR AND VACUUM

Power requirements for compressors and vacuum pumps must be considered when wiring plans are made. Both compressed air and vacuum are frequently used in laboratories. Sometimes both can be piped to more than one laboratory room from a central location. Air pressure will usually be sufficient for laboratory applications, but vacuum may not always be, in which case a separate vacuum pump would be required.

Even a small compressor tends to be very noisy and for that reason should not be in the laboratory proper. Rough plumbing will carry air to the laboratory area. Since the moisture in compressed air causes rusting, the use of rust-resistant pipe will substantially lengthen the life of the air filters. Vacuum pumps, on the other hand, are usually not too noisy to be in the laboratory, although the pump selected should be checked for noise before such a decision is made.

VENTILATION

A well-ventilated laboratory is still not as common as it ought to be. While local building codes may not require it, any laboratory should have a well-designed forced air ventilation system. It not only promotes worker comfort but has a strong effect on safety.

Many designers of ventilation systems treat a laboratory like an office, calculating the volume of the area and using standard formulas for planning the system. Under some conditions this works very well and conserves much energy for both heating and cooling. The typical ventilation system recirculates a major portion of the air, however, and in many laboratories this is not acceptable. Even though work creating significant amounts of hazardous or unpleasant fumes is performed in hoods, many fumes may still come from the work bench. In addition, non-hazardous amounts of corrosive fumes will in time have their effect on metals.

The ideal laboratory ventilation system is the once-through type, where all air is exhausted to the outside after a one-way trip through the area. It is expensive because it raises both heating and cooling requirements, but it will still pay off. In one laboratory with such a system, small amounts of corrosive vapors, far from enough to be called hazardous, would occasionally develop. After five years, an expensive analytical balance sitting in the middle of the room still showed no signs of corrosion. Balances of the same age in a separate weighing room of a state university laboratory showed definite marks of corrosion on internal parts. This building, though, had conventional ventilation.

No proposal for a laboratory ventilation system should be requested without a thorough study of the work to be performed. A once-through system may not always be required. The manager of a research laboratory, working closely with a designer, discovered that 50% recirculation would be permissible in their new building. While this is even less than in offices and stores, it proved to be adequate for their type of work. He would not recommend this ratio for other laboratories without a careful study.

When a recirculating system is to serve non-laboratory areas as well as laboratories in the same building, great care must be taken. In the chemistry building of a California community college, the same system served lecture rooms, offices, and laboratories. Despite the use of fume hoods for unpleasant operations, freshman chemistry classes still created fumes which would at times permeate the whole building. While considerably more expensive, a separate system for the laboratories would have been advisable. A split system in this case would have had an additional advantage. During the night, while ventilation for the rest of the building was shut off, it could have been operated at low speed in the laboratories to prevent build-up of fumes. This would have increased safety and minimized equipment corrosion as well.

In another college laboratory building, air intake for ventilation and fume hood exhaust ducts were placed too close together on the roof. Under certain wind conditions, unpleasant odors would fill the whole building, causing many caustic comments.

Close cooperation between laboratory operator and designer can sometimes bring about savings along with good results. Once-through ventilation was requested for a laboratory to be installed in a 30x30 foot building. After a careful study of the plans, the designer suggested that a separate exhaust fan could be eliminated at significant savings. He laid out a carefully planned system of intake and exhaust vents in the ceiling. The exhaust air went to the attic, which had large vents to the outside. The ventilation was good and, in addition, the attic area was heated in winter and cooled in summer.

A good idea that did not work out was found in a laboratory where flammables were to be handled. The climate was such that air conditioning was not felt to be needed. The designer came up with an excellent ventilation system but, unfortunately, he placed the air intake on the west side of the roof, practically overhanging the wall. During the summer, the prevailing wind from the northwest would sweep across the sun-baked parking lot and then contact the hot concrete wall of the building before entering. This problem could have been avoided had the intake been placed on the north side of the roof where the temperature was much lower.

In a cosmetic laboratory, a recirculating system would have been adequate except for one very special condition. The "noses" had to perform their fragrance evaluations in a clean atmosphere. This called for once-through ventilation and included placing drying ovens under a simple hood extending part way down from the ceiling. The result was excellent.

HEATING AND COOLING

A good heating and air conditioning system provides more than creature comfort for those working in a laboratory. Conditions of reasonably constant temperature and humidity are important for the proper performance of many laboratory operations.

Conventional hot-air furnaces, similar to those found in homes, are often used in laboratories. A university campus or an industrial plant may have steam available, which can be conveniently put to use for heating.

Air conditioners are frequently combined with heaters. A convenient place must be found for the compressor, generally outdoors. Heat pumps have been used a great deal in recent years with satisfactory results. In very dry climates, laboratories often install evaporative coolers because of their reasonable cost of installation and operation. Their one drawback, sometimes serious, is that the inside air tends to become uncomfortably humid.

Estimating the amount of heat, cooling, or air flow required is a job for a skilled professional. Such an expert should also be called in when it seems desirable to extend a system to areas other than those for which it was originally designed.

4

Laboratory Safety

A safe laboratory is the result of both good design and proper work rules. The laboratory operator, as the only person fully aware of the work to be performed, must be involved in all safety planning. He will be the one who can supply the safety experts with the information they need.

SAFETY AWARENESS

Much has been done to improve laboratory safety over the years. Two examples from the author's early experience can illustrate the lack of precautionary measures as late as the 1940's.

The first is the story of the high school physics teacher who came down with a strange illness eventually diagnosed as mercury poisoning. The cause? Over the years, minute amounts of this metal had been spilled in the poorly ventilated lecture room. The students suffered no ill effects since they spent so little time there. Fortunately, the teacher recovered, the room was completely refurbished, and strict rules were introduced concerning the handling of mercury.

The second example, also involving mercury, comes from an

otherwise well-operated brewery laboratory. Periodically, dirty mercury would be taken from tank manometers to the laboratory for cleaning. It was transported in beer bottles, which had to be handled with great care to avoid breakage. In the laboratory, the mercury was placed in a large porcelain dish and treated with dilute acid, followed by several water rinses. When the acid and water were decanted into the sink, a little mercury would go there, too. Final drying of the mercury was accomplished by stirring it with highly absorbent filter paper, hand held, which then went into the garbage. Finally, the mercury would be filtered through glass wool into clean beer bottles for return to the plant. The used glass wool was discarded as regular garbage.

When questioned about the safety of such procedures, the chief chemist had a quick answer: "We have done it this way for over 20 years and never had a problem." Such handling of mercury would be unthinkable today.

There are other more recent examples though. In one laboratory, flammable solvents had been used in such small quantities that they required no special storage. As their use of solvents increased, however, there were eventually five-gallon cans of ether casually stored in one corner of a room. When a routine inspection by the fire department uncovered this hazard, the laboratory was forced to install proper storage.

In a testing laboratory, the regular fume hood was used for performing occasional perchloric acid digestions. As the number of such procedures grew, no special precautions were taken. Finally a concerned analyst spoke up and insisted that an appropriate hood be installed for such work. The management grumbled at the additional expense but, fortunately, recognized the need before any accidents had occurred.

In another laboratory, the chemist was annoyed by the fact that so much ether evaporated from containers in hot weather. He decided to store the ether in a regular household refrigerator made "safe," he thought, by disabling the light switch at the door.

Eye protection was rarely practiced in older laboratories except when particularly hazardous work was carried out. Today such protection is mandatory for practically all laboratory operations.

Older fume hoods were anything but efficient in many cases. One laboratory had three hoods but only one worked well, and workers sometimes had to wait for it to become free. Even the good hood was never checked for face velocity. It just operated better than the others. The management showed little interest in repairing or checking the hoods as long as they caused no serious work delays.

Older laboratory workers could no doubt add many examples of their own to the list of unsafe practices. Modern laboratory planners and operators, however, are more aware of safety problems, and rules regarding safety have been made much more strict. It is up to the laboratory operator to become throughly familiar with such rules in order to organize and operate a safe laboratory.

SOURCES OF SAFETY INFORMATION

Information on laboratory safety is available from many places. The public library usually has material on federal, state, and local safety regulations. The *Index to Books in Print,* also available at libraries, lists all material published by subject, author, and title. Scientific book publishers will be glad to add new names to their mailing lists for announcements of new publications on safety.

A good place to start looking for information pertaining to a laboratory in the planning stages is a college or university library, which will have a good collection of books on safety. Some are general, others quite specific. A careful review will show which books should be considered for purchase for the new laboratory library. The publication dates of such books must be noted, since rules and regulations which may be referred to can change. Permissible exposure limits to chemicals, for instance, are frequently updated.

Professional organizations, such as the American Chemical Society, have published information on safety. Professional journals often carry safety articles, including stories about laboratory accidents of types not formerly experienced.

Material Safety Data Sheets (OSHA Form 20, see Figure 1) are available from suppliers for every chemical sold. These give valuable information of a specific nature and should be kept on file. A phone call to the nearest office of the Occupational Safety and Health Administration (OSHA) of the U.S. Department of Labor will bring forth more information on safety, both oral and written.

The local building department will have information on local requirements, and the local fire department will give helpful advice and call attention to special regional restrictions.

Insurance companies also frequently have experts on the staff from whom valuable knowledge may be obtained. Before writing a policy, they may want to go over the laboratory's operation with a fine tooth comb.

DEALING WITH AUTHORITIES

After informing himself on regulations and other safety information, the laboratory operator will be ready to meet and discuss plans with the authorities. To many, this is considered at best a nuisance to be avoided if at all possible. Some think, for instance, that a call to OSHA for information will alert this agency to the fact that a laboratory is there and will be followed up by inspection. Likewise, they fear, a call to the local building department

U.S. DEPARTMENT OF LABOR
Occupational Safety and Health Administration

Form Approved
OMB No. 44-R1387

MATERIAL SAFETY DATA SHEET

Required under USDL Safety and Health Regulations for Ship Repairing, Shipbuilding, and Shipbreaking (29 CFR 1915, 1916, 1917)

SECTION I

MANUFACTURER'S NAME	EMERGENCY TELEPHONE NO.
ADDRESS *(Number, Street, City, State, and ZIP Code)*	
CHEMICAL NAME AND SYNONYMS	TRADE NAME AND SYNONYMS
CHEMICAL FAMILY	FORMULA

SECTION II - HAZARDOUS INGREDIENTS

PAINTS, PRESERVATIVES, & SOLVENTS	%	TLV (Units)	ALLOYS AND METALLIC COATINGS	%	TLV (Units)
PIGMENTS			BASE METAL		
CATALYST			ALLOYS		
VEHICLE			METALLIC COATINGS		
SOLVENTS			FILLER METAL PLUS COATING OR CORE FLUX		
ADDITIVES			OTHERS		
OTHERS					

HAZARDOUS MIXTURES OF OTHER LIQUIDS, SOLIDS, OR GASES	%	TLV (Units)

SECTION III - PHYSICAL DATA

BOILING POINT (°F.)		SPECIFIC GRAVITY (H_2O=1)	
VAPOR PRESSURE (mm Hg.)		PERCENT, VOLATILE BY VOLUME (%)	
VAPOR DENSITY (AIR=1)		EVAPORATION RATE (________ =1)	
SOLUBILITY IN WATER			
APPEARANCE AND ODOR			

SECTION IV - FIRE AND EXPLOSION HAZARD DATA

FLASH POINT (Method used)	FLAMMABLE LIMITS	Lel	Uel
EXTINGUISHING MEDIA			
SPECIAL FIRE FIGHTING PROCEDURES			
UNUSUAL FIRE AND EXPLOSION HAZARDS			

Figure 1: Material Safety Data Sheet

SECTION V - HEALTH HAZARD DATA
THRESHOLD LIMIT VALUE
EFFECTS OF OVEREXPOSURE
EMERGENCY AND FIRST AID PROCEDURES

SECTION VI - REACTIVITY DATA			
STABILITY	UNSTABLE		CONDITIONS TO AVOID
	STABLE		
INCOMPATABILITY *(Materials to avoid)*			
HAZARDOUS DECOMPOSITION PRODUCTS			
HAZARDOUS POLYMERIZATION	MAY OCCUR		CONDITIONS TO AVOID
	WILL NOT OCCUR		

SECTION VII - SPILL OR LEAK PROCEDURES
STEPS TO BE TAKEN IN CASE MATERIAL IS RELEASED OR SPILLED
WASTE DISPOSAL METHOD

SECTION VIII - SPECIAL PROTECTION INFORMATION		
RESPIRATORY PROTECTION *(Specify type)*		
VENTILATION	LOCAL EXHAUST	SPECIAL
	MECHANICAL *(General)*	OTHER
PROTECTIVE GLOVES		EYE PROTECTION
OTHER PROTECTIVE EQUIPMENT		

SECTION IX - SPECIAL PRECAUTIONS
PRECAUTIONS TO BE TAKEN IN HANDLING AND STORING
OTHER PRECAUTIONS

Figure 1: (continued)

about a small laboratory installed in a corner of a room could also put its owner on record as the operator of a laboratory and force compliance with all kinds of "silly" rules. File cabinets full of accident reports back up the fact that these rules are not so silly. In fact, discussions with such agencies may make even a highly experienced laboratory operator aware of hazards he did not know existed.

Unfortunately, in the minds of many regulators the word *laboratory* may conjure up visions of fires, explosions, and highly toxic materials. For that reason, a laboratory planner should be prepared to present the proposed operation in great detail and to answer all questions, even those which may not seem relevant. Any work planned for the future should also be discussed, since the added cost of accommodating for it in the beginning may be just a fraction of what would have to be paid for later modifications. All pertinent facts must be revealed; a minor omission could make a big difference in laboratory safety. A laboratory operator who has studied the various rules and regulations with care will have no problem handling such discussions.

It should be noted that non-compliance with safety rules would be held strongly against a laboratory should there be a lawsuit as a result of an accident.

VENTILATION SYSTEMS

Laboratory ventilation and fume hoods are discussed in detail elsewhere in this book. Here the concern is for the safety aspects.

The ventilation system must be able to carry away hazardous fumes that may form during normal work. Monitoring the laboratory atmosphere for such fumes while work is in progress is not only advisable, but may even be required by law. Monitoring must be repeated whenever new fume-producing procedures are introduced or any time modifications are made to the ventilation system. Local health authorities should be contacted about this. Some laboratories purchase fume detectors from safety equipment suppliers.

Fume hoods must be of a type suitable for the service they are intended to perform. For many applications, minimum face velocity is specified by regulations. An installer should always check the velocity when a new hood is placed in operation. It should be rechecked whenever any modification is made to the exhaust system. It is up to the laboratory operator to make certain that a hood is not put to new uses for which it was not designed.

FLAMMABLE MATERIALS

Storage and use of flammable materials are often not given sufficient consideration. A safe storage cabinet is obligatory if more than very small amounts of flammables are to be used. These cabinets come in many sizes and shapes, including wall-hung models. There are even types large enough to hold a 55-gallon drum. Cabinets conforming to official specifications are not cheap, but a planner should not be tempted to get by with one that seems "just as good." In addition, the cabinet selected should be large enough to take care of future needs.

Some larger institutions build special explosion-proof rooms for storage of flammables. These must conform to all official rules with respect to design, construction, and location.

Work with flammables should be confined to areas where other work that could cause ignition is not performed. This practice must be carefully policed. Extra ventilation may be needed for such an area. Knowing that most flammable fumes are heavier than air, one laboratory installed an explosion-proof exhaust fan at floor level below the bench where flammables were used. To make the area for working with flammables easier to identify, it could be marked off with red lines on the floor.

Laboratories working frequently with flammables, such as in extraction processes, may set aside a separate room for such work for maximum safety. Ideally, this room should have explosion-proof electrical equipment and special ventilation. Rigid work rules should be enforced, such as requiring workers to leave matches and lighters on a shelf outside before entering. Installation of conductive flooring should be considered, and steel tools should be prohibited at times when work is in progress.

CHEMICAL SPILLS

Chemical spills in the laboratory may be hazardous, damaging to flooring and furniture finishes, or simply messy to clean up. Laboratory equipment houses and distributors of safety equipment carry a variety of clean-up kits made for specific types of spills, depending on the types of chemicals handled. Generally, the kits contain absorbents or neutralizers, or a combination of the two, plus instructions for use. Kits should be stored as close as possible to where spills are likely to occur and not in a remote storeroom. Instructions must be kept with the kits, not in a file cabinet. Quick action in case of a spill is important in reducing both hazards and damage.

The most common spill on a laboratory floor is water, not generally considered hazardous. It may, however, create hazardous conditions by making the floor slippery. It is dangerous to walk on a wet floor with certain types of soles. A wet area should always be blocked off until it is dry. Mop and bucket should always be available close by, not in a janitor's closet down the hall. String mops are superior to sponge mops for getting the floor dry.

Spills of oily materials will make most floors feel like an ice rink. A simple wipe-up will leave enough of a film to maintain slippery conditions. Cleaning must be done with detergent and water, followed by rinsing.

SAFETY SHOWER AND EYEWASH STATION

A safety shower is required whenever there is even the slightest possibility that clothing could catch fire or where hazardous chemicals could be spilled on skin or clothing. The use of safety blankets, though still available, is no longer recommended by some safety experts. The synthetic fabrics used in today's clothing, they claim, can melt from the heat and stick to the skin, thereby creating a secondary problem. In case of spills, of course, blankets are of no use.

While there may be places where a shower is not needed, one or more eyewash stations will almost always be required. Some laboratories, in fact, provide a small eyewash device at every sink. Combination shower and eyewash units are also available. Home-made devices are not satisfactory. A standard safety shower gives off a drenching stream of water of much higher volume than a home-type shower. An eyewash emits a large volume of water at a gentle pressure to prevent eye injury.

It must be possible to reach a safety shower within ten seconds from any area where hazardous conditions exist.

CHEMICAL STORAGE

Storage of flammables has been discussed separately, since these materials present special hazards, but there are many other materials that must be stored with special care. These may be toxic or corrosive, or they may be capable of entering into violent reactions if mixed in case of breakage or spills. Some may give off hazardous fumes even if stored in nominally closed containers.

A few general rules for storing chemicals can be listed here:

> Any chemical that may give off hazardous fumes must be stored in a separate and well ventilated area. Extra precautions must be

taken if fumes from different chemicals are likely to react with each other.

- Some chemicals may need refrigerated storage because of high vapor pressure. If the vapors are flammable, an explosion-proof refrigerator is mandatory, despite its high cost.
- Large containers of chemicals should be stored as close to the floor as possible, never on a high shelf.
- Storage shelves should not be so deep that one has to reach behind other containers to get the desired one.
- Chemicals likely to enter into violent reactions should not be stored close together, since there is always a possiblity of container breakage.

Additional information may be had on storage of chemicals from several sources. Material Safety Data Sheets, for example, have specific instructions with regard to storage. The local fire inspector will have good suggestions, and much can be found in reference books. Above all, the laboratory operator should throughly familiarize himself with all chemicals to be used in order to develop a safe storage system.

When chemicals are purchased in large quantities (as is often the case in universities), the individual containers will arrive in sturdy cardboard or foam plastic boxes which give good protection against physical damage. They may be left in the boxes and removed when needed to stock the shelves. Of course, different chemicals must not be kept in the same box.

An efficient storage system may not always be a safe one. In a large testing laboratory, all reagents had been arranged in alphabetical order on shelves, making it quick to find any given material among the hundreds of bottles and jars. This convenient arrangement, however, caused some large containers of hazardous materials to end up on high shelves, and reagents likely to react with each other sometimes found themselves side by side alphabetically. A fire inspection officer ordered the reagents stored in a safer manner. Although it now took longer to find a given reagent, safety was greatly improved.

FIRE PROTECTION

In a laboratory, fire hazard may vary from minimal to severe. Proper protective measures, though costly, are worth the price. A laboratory must meet local fire protection standards for industry, which are usually more restrictive than those for a home. While a fire department official is not a chemist, given the proper input he will come up with good suggestions. Providing him with the necessary information is an important job for the laboratory planner.

Problems do occur when a laboratory is installed in an existing building on a small scale, often without a building permit, or when an existing laboratory is expanded in a so-called minor way. Such situations are common in industry. A prudent laboratory operator should demand complete compliance with rules in such cases despite cost, which could be high, particularly when old mistakes have to be corrected.

If sprinklers are required, or if the sprinkler system is to be modified, such work should only be performed by a company certified to do it. Fire extinguishers are almost always required. The number of them, their placement, and the types needed are matters to be discussed with fire officials. Extinguishers should be regularly serviced by qualified specialists.

Fire drills are very much in order in most laboratories. When the signal is given, all workers should shut down their equipment and leave in a quick and orderly manner. Some employees may be assigned to fire extinguishers or other fire fighting equipment. The speed with which this is accomplished is important and may be timed with a stopwatch. After the drill, a supervisor may want to check on how well the equipment was secured in order to minimize hazards and damage. Fire departments are happy to instruct personnel in fire safety or to put on demonstrations.

EARTHQUAKE PREPAREDNESS

Although earthquakes are more frequent in some areas, such as California, they can occur almost anywhere. Places with a history of quakes have strict regulations regarding precautions, but occasionally a jolt in some place like New York reminds us that a laboratory planner should take them into account.

There are two important safety measures that can be taken at very low cost. The first is to equip shelves with guard rails to keep reagent bottles from falling. The height of such rails should be adapted to the size of the containers. This is a requirement in California. The second is to fasten tall objects to the wall, a simple and inexpensive procedure. With bookcases, for example, the books may fall out but the case will stay in position and not tumble down with the whole load.

Earthquakes may also jam doors, another good reason for having more than one door leading from a laboratory.

Procedures to follow in case of an earthquake or a fire should be posted and reviewed with laboratory personnel from time to time.

MECHANICAL HAZARDS

Most mechanical hazards in the laboratory can be avoided by good planning. There must be no obstructions in traffic areas, particularly those needed for rapid evacuation in case of emergency. Wide hallways may look like good parking places for movable equipment not currently in use. They may also seem convenient for storage of reagents and supplies that temporarily do not fit into regular storage places. The use of passageways for storage, even on a temporary basis, is very unsafe.

Waste containers and stools are frequent occupants of laboratory aisles, where they may cause congestion. One answer to this is to provide a sufficient number of knee-holes in the work benches to hold them.

Equipment

Modern equipment is relatively free from mechanical hazards. For instance, there is always a shield covering drive belts so that they cannot catch fingers or clothing. Nevertheless, a prudent planner will still place equipment with moving parts in a location where contact with workers is unlikely.

Any piece of equipment under pressure is a potential hazard and must be properly shielded. Shields of clear plastic, properly anchored, will often suffice. Vacuum has its problems too. More than one glass flask with an almost imperceptible fault has imploded under vacuum, with an effect much the same as an explosion. Here, too, shielding is recommended.

Doors

Doors present certain hazards, too, since a collision is likely to take place when two people approach a door from opposite directions at the same time. Even a small window in every door makes it possible to observe activities on the other side and prevent such problems. This may not be possible in those few cases where fire resistant doors are called for.

Swinging doors are often considered practical, since a person can walk through with both hands full. Many types, though, will swing back with a vengeance, causing problems for the next person entering.

UTILITY FAILURES

Most laboratories depend upon a steady supply of electric power, water, and gas. Very unsafe conditions may develop if any of these should fail, so precautionary measures must be taken.

Electricity

Electrical failures are by far the most common. The most important safety measure is the installation of emergency lights, which will go on as soon as the power fails. Their batteries will automatically be recharged after the power comes back. They should be installed even in laboratories where work is normally performed only during the day. Such units are fairly expensive, but smaller ones made for home use are now available in hardware stores. A check should be made with local authorities as to whether these are permissible for industrial use.

Water Pressure

Failure of water pressure can spell disaster in laboratories where some operations, such as distillations, are dependent on water. Ideally, an audible alarm would indicate the lack of water pressure. Laboratory supply houses sell water flow indicators that can be placed in series with the water line to equipment. When water is flowing, a brightly colored ball in a clear tube will move or a small propellor will turn. These can be observed from a considerable distance by watchful laboratory personnel.

Gas

A gas failure is rare but could present a serious hazard if it should occur. Unless turned off, a gas-powered piece of equipment will pour unburned gas into the room after service has been restored.

How to Handle Utility Failures

The simplest way to cope with such potential hazards is to develop a set of instructions about what to do in case of failures. These could be posted, preferably in a place where they can be read with the emergency light. As soon as a failure occurs, laboratory personnel should immediately turn off all affected equipment. In case of an electrical failure, attention should be paid to samples that may be giving off hazardous fumes from the hood.

Most electrical failures are caused by external problems, such as a malfunctioning transformer, a thunderstorm, or a car hitting a power pole. Gas and water pressure failures are often caused by problems within the building itself, such as somebody shutting off a valve in order to make a repair without notifying those concerned. A system must be established whereby all affected by such a shut-down will be informed well in advance.

PERSONAL PROTECTION

Cleanliness comes very close to godliness in the laboratory. A worker always washes hands before eating or smoking, and many times in between. One thoughtful laboratory operator placed bottles of liquid skin cleanser and hand lotion next to all sinks. Apart from personal cleanliness, there are other important forms of protection for the worker. Most notable among these is eye protection.

Safety Glasses

Eye protection is mandatory for almost all types of laboratory work. A wide selection of safety glasses and goggles is available from suppliers of safety equipment, many especially designed for specific purposes. There are official specifications for such glasses, and it is unwise to buy any which are claimed to be "as good as" the ones guaranteed to meet those specifications.

How much protection does one get from regular prescription glasses? Actually quite a bit. Current laws require that they meet impact resistance requirements, except in case of certain special prescriptions. Such requirements, however, are less rigid than those for safety glasses. Even the currently popular large lens glasses lack side shields. Clip-on side shields could make regular prescription glasses perform as well as many types of safety glasses.

Contact lenses, on the other hand, must never be worn when there is even the slightest chance that hazardous fumes could reach the eyes. The fumes could work their way between the contact lens and the surface of the eye.

Some laboratory operations require the use of goggles, which effectively protect the eyes from splashes or missiles from all sides, or even complete face shields. The latter are uncomfortable to wear for extended periods of time, but they do give superior protection for jobs such as making concentrated solutions of sodium hydroxide. Many goggles are designed to be worn over regular glasses and are reasonably comfortable.

Lab Coats

Lab coats do more than protect clothing. They also protect the wearer against chemical spills and are easily removed if a spill should take place. Cotton fabrics are both the most comfortable and the most resistant to chemical spills. For certain operations, disposable protective clothing may be required.

Foot Protection

The last drop from a liquid transfer always seems to find its way to a worker's toes. The operator of a laboratory where highly corrosive chemicals were handled declared that only solid top shoes could be worn. This unpopular edict was reinforced by a display of his old shoes with many spill marks on the toes.

Food and Drink

Food in most cases should not be in the laboratory, not even candy bars. There is always a chance that food products could become contaminated by chemicals on the work bench or on unwashed hands. There should be designated areas where food can be consumed.

Laboratory workers seem addicted to coffee, and in a typical facility coffee cups can be found at every work station. 250ml beakers are almost ideal for coffee, but their use should be discouraged. It is too easy to pick up the wrong beaker on a crowded work bench. One supervisor gained much popularity by presenting his staff with a set of mugs for Christmas. Despite a shortage of space, someone found a safe and sanitary place for storing them. Of course, all food and drink must be banned from areas where toxic materials are handled.

Smoking

Smoking must not be permitted in areas where flammables or toxic substances are used. The laboratory operator should provide for a safe smoking area, generally one where food may also be consumed. In work areas where smoking is permitted, inexpensive ashtrays should be used. They are more practical than Petri dishes.

Hair Protection

Long, flowing hair may be glamorous, but it does not belong in the laboratory, where it may catch fire or get into moving equipment. Such hair should be rolled up during working hours. Universities are very particular about this, too, but one professor half-jokingly mentioned his concern over long beards in the chemistry classes, which might get too close to a Bunsen burner.

SAFETY SIGNS

Safety signs of approved types, available from laboratory supply houses and safety equipment dealers, should be posted in appropriate spots. The door leading out of the laboratory should be marked EXIT, while the door to a back room should be marked NO EXIT. The location of a fire extinguisher must be clearly marked. Signs are available for every type of hazard. Homemade signs not conforming to official standards should not be considered.

CONCLUSION

Every laboratory is different. Standard safety features when applied to a given situation may be insufficient in some cases, superfluous in others. There will also be times when work is being done with materials or procedures not covered in available literature. Here the laboratory operator must use a combination of experience, knowledge, and common sense. The last of these may often be the most important. Planning a laboratory and making it operate safely thus becomes a custom procedure.

5

Pollution Control and Waste Disposal

Until recent years, little attention was given to pollution control or waste disposal except when highly toxic or flammable materials were involved. Undesirable fumes, regardless of nature, were simply sent up the stack until neighbors complained about obnoxious odors. The remedy? Just make the stack higher. Anything liquid went down the sink to be flushed away with water and thereby considered "diluted." Solid waste of all kinds went into the garbage and got sent to the local dump. Hard to dispose of chemical mixtures were often taken to an open lot behind the laboratory and dumped on the ground.

WASTE CONTROL REGULATIONS

Pollution control and waste disposal are now under strict regulation at federal, state, and local levels. Enforcement of these regulations has also become more effective. Increased public awareness and media coverage make it difficult for today's polluter to hide behind a screen of ignorance.

Local regulations are sometimes more restrictive than federal or state laws and should be carefully studied by a laboratory planner. City and county governments in industrialized areas are increasingly creating new positions in this field for inspectors, analysts, and related experts. Universities, too, have added personnel to coordinate and direct disposal of their laboratory waste.

Even though the quantity of pollutant or hazardous material from a laboratory may be small, laws often concern themselves with quality rather than quantity. Thus the small amount of sewage from a laboratory may fall under the same rules as the large flow from an industrial plant as far as pollutant concentration is concerned. Throwing a little prohibited material into a garbage bin may be as unlawful as disposing of larger quantities. In fact, even empty containers which once held hazardous chemicals need special disposal. Sometimes inspectors may not seem to be too concerned about a laboratory since they have to spend so much time on the larger dischargers, but this is no reason for neglecting the rules.

AIR POLLUTION

Few laboratories have true air pollution problems, but the possibility should always be considered. Some may give off toxic fumes, which are regulated by law and therefore must be controlled. Others may produce fumes which, though not toxic in the eyes of the law, may be foul-smelling or irritating. While outside the official reach of regulations, neighbors may eventually become irritated to the point of instigating legal action.

An example of this type of problem was seen in a laboratory performing macro-Kjeldahl analyses on a large scale. This procedure sent sulfuric acid fumes into the atmosphere through a stack on top of the building. Usually the fumes did go away with no problems, but workers in a nearby business building complained about eye irritation when the wind was right. There were also extensive corrosive effects on the building's roof. Both problems were solved by scrubbing the fumes with a fine water spray, but this sent far too much acid into the city sewer. Finally, sodium carbonate solution was injected into the water used for scrubbing, a treatment which satisfied the local waste water inspector.

Flammable vapors in the amounts emitted from a laboratory are rarely of much concern, but it is a good idea to make some estimates. Usually these vapors are quickly dispersed in the atmosphere to the point where they cause no problem. Care should be taken that there are no sources of ignition close to the top of the exhaust stack.

LIQUID WASTE

No liquid waste can be disposed of directly into the environment or into a storm sewer system. Unless collected and handed over to a waste disposal company, it goes into the sanitary sewer. There are specific rules concerning the quality of what may be disposed of in this manner. Materials that are routinely sent to the sewer from a home may not be permissible from a commercial source. Normally accepted waste water rules are often made more restrictive by local ordinance.

Waste water rules have pH limits, a common range being between 6 and 10. There are also limits for fats and oils, solvents, heavy metals, and a variety of other compounds and ions. The fact that a compound with possible toxic or otherwise undesirable properties is not on the list does not mean it is permissible. Such a matter should be discussed with the proper authorities. The discharged water may also have to pass a test for toxicity to aquatic animals. As one frustrated manager of a chemical plant put it: "We can no longer put anything but pure tap water into the sewer!" Of course, it is not really that bad, but some of the requirements often come as a surprise.

Waste water rules sometimes require that water conform to the limitations at all times. Under such rules, surges of non-conforming sewage are not permitted. Tests run on composite samples over a 24-hour period, for example, will then not be acceptable.

SOLID WASTE

Whatever goes to the local sanitary dump will eventually find its way into the environment. For this reason, materials that are to be disposed of as garbage from the laboratory should be carefully scrutinized. There are those who feel that traces of hazardous materials when mixed with large amounts of regular garbage will somehow disappear. They will not. Worse yet, some materials could react when combined and possibly generate enough heat to cause fire.

Broken glassware free from hazardous chemicals, paper towels, empty containers that have not held toxic, flammable, or highly reactive chemicals are all good candidates for the garbage container. For any other waste material, the laboratory operator has to make decisions. Filter paper, for instance, could contain residue that should not be disposed of as garbage. Anything that contains heavy metals or ionic species that may be termed hazardous must be given special disposal. The same is true for both organic and inorganic materials from which natural processes, such as leaching, could extract hazardous substances.

WASTE COLLECTION AND DISPOSAL

The laboratory operator must make a careful examination of all wastes that will be generated and, from this, work up a waste disposal system. Some wastes may be compatible and could be disposed of together. Others could react and thus cause problems. Flammables must be given special attention. Certain biological wastes may be very hazardous even in small quantities. Special rules apply to radioactive materials, even in the small amounts used for investigative purposes.

Disposal Containers

In the laboratory, wastes are collected in suitable containers placed in convenient locations. Kneeholes in work benches are good for this purpose, since the containers are then out of the way. Waste containers must be compatible with what is put into them. Acidic materials, for instance, may quickly damage steel buckets. Some plastics may not stand up to solvents placed in them, and so on. Flammables must go into approved safety cans. Dry waste should be put into self-closing containers designed to snuff out a possible fire. Lining solid waste containers with plastic bags is highly recommended.

All containers must be clearly marked as to contents. Color coding with wide adhesive tape is one way to do this. Special warnings should be placed on containers for particularly hazardous materials. A typical example of poor waste management reported in the press concerned an unmarked waste container in a university laboratory. When the container showed signs of heat-producing reaction and began to leak, a professional clean-up team was called in. Although only four gallons of material was involved, 27 different chemicals were identified in the mixture!

Waste Storage

It is important to keep the amount of potentially hazardous waste in the laboratory to a minimum at all times. Periodically, contents of the laboratory waste containers will be transferred to appropriate containers for final disposal. These must be stored in a safe location, often outdoors, while awaiting pick-up by a disposal service. There will be local restrictions for such storage. A locked storage area may be needed, for example, to prevent unauthorized access to hazardous materials. The fire department may set strict limits as to how much flammable material may be present. All containers must be marked with contents, and the storage area will no doubt require warning signs.

Waste Disposal Services

There are now many companies specializing in waste disposal.

Local ones are listed in the yellow pages of the telephone directory. The laboratory operator should contact more than one for suggestions and estimates. In such discussions, a disposal service will ask detailed questions as to the exact nature of the waste and quantities involved. An agreement will be made about the types of shipping containers to be used. These are nonreturnable. The disposal service may sell suitable containers and require these for certain types of waste. In other cases, the laboratory may provide its own. If drums are needed, they are available from companies engaged in drum reconditioning. Manufacturers of chemicals may have used drums available at reasonable prices, but since these will contain residues of their former contents, a check for compatibility must be made before using them.

IN-HOUSE WASTE REDUCTION

The cost of waste disposal is significant. Unfortunately, it is often overlooked in the laboratory budget, a fact which has encouraged unlawful disposal. The laboratory operator should face this matter head-on, since delaying action will just make the problems multiply. One possible choice for laboratories producing large amounts of hazardous or otherwise unacceptable waste is to look into ways to reduce the volume of such waste or modify its composition to where it is acceptable for normal disposal. Strongly acidic or alkaline solutions, for instance, may sometimes become acceptable as regular waste water after neutralization.

Some chemicals may be reacted to produce compounds that are acceptable for regular disposal, but this should be a matter of discussion with appropriate authorities. Solutions of non-volatile but hazardous materials may be evaporated to low volume, while volatile solvents could be reclaimed by distillation. Although organics and biological materials may be incinerated, just getting a permit for operating an incinerator is a complex and time-consuming procedure. Unfortunately, both equipment and labor for waste reduction are costly, so the resultant savings may be questionable.

DISPOSAL OF OLD CHEMICALS

Any laboratory that has been in operation for some years will have a collection of old chemicals purchased for past projects, which may have been completed or just abandoned. Even though they take up valuable storage space, there is a feeling that they may be useful some day. Over time, however, some reagents may have picked up enough moisture to be transformed from free-flowing powders to rock-hard substances containing unknown amounts of water of crystallization. Others may have hydrolyzed to a point where the original analysis is meaningless. Heat and light may have caused some to break down, and a few may have become dangerous. For example, a half-filled can of ethyl ether that has been sitting on the shelf for an extended period of time is best handled like a potential bomb. Periodic inventory of chemicals stored in the laboratory is therefore very much in order.

Before demolishing the old chemistry building on its campus, a university called in a disposal service to get rid of old chemicals stored here and there. While this was not cheap, it was felt to be the safest and most efficient way to handle what could have been a potentially hazardous situation.

A community college had a similar problem when a public-spirited industrial company donated its surplus reagents to the chemistry department. Some were indeed useful, but others were not and never would be. They just sat around for years in boxes and on shelves, posing potential hazards and taking up space that could have been put to better use.

DIFFERENT LABORATORIES, DIFFERENT PROBLEMS

Since pollution control and waste disposal problems vary greatly from one laboratory to another, they have to be handled on a custom basis. The laboratory operator will no doubt be an expert on the materials used but not on their disposal. It is important for him to stay in touch with people familiar with disposal and provide them with detailed information.

Industrial laboratories often handle a relatively limited number of chemicals. Since many of these same chemicals are also used on a manufacturing scale, their disposal will have already been planned for by plant personnel. Disposal of chemicals used only in the laboratory, however, for investigative and development purposes must also be considered, as well as reagents used for testing.

Laboratories handling biologicals have their own special problems. They often have to dispose of small but significant amounts of materials that may be very hazardous. Every laboratory procedure must then be scrutinized with this in mind.

University and college laboratories must dispose of a great variety of chemicals. Some may even be types for which hazards have yet to be well defined. Setting up a disposal program then becomes a complex procedure, calling for cooperation by all concerned. These laboratories also keep a large number of chemicals on their stockroom shelves, making periodic inventory time-consuming.

THE ILLUSION OF DILUTION

How much is too much? This question comes up frequently in discussions about waste disposal. After all, "a little" hazardous material should not cause much damage or be in violation of regulations. It will simply disappear if sufficiently well mixed with a large quantity of harmless materials. Such thinking might be called "the illusion of dilution." A good example comes, not from a laboratory, but from a manufacturer of semiconductors.

At the semiconductor plant, a small amount of hydrofluoric acid was used in the processing. After neutralization, it was discharged into the large flow of water from the plant. When waste water in the area near the plant showed fluoride levels well in excess of the permissible 5ppm, the manufacturer became suspect. Refusing to believe that he could cause such contamination, he called in a consultant, who calculated that the water from this plant could contain as much as 60 ppm fluoride. This still sounded impossible to the manufacturer, but analysis confirmed the calculated result. Steps were soon taken to correct the situation.

It is always surprising to find how even very small quantities of some chemicals can create contamination or waste disposal problems. As a result, such matters should be given high priority when planning a new or modified laboratory, purchasing new equipment, or introducing new procedures.

6

Floors, Walls, and Ceilings

In selecting the finishing materials for floors, walls, and ceilings, utility and cost are major considerations. The type of work to be performed in the laboratory, however, will be the determining factor in making the choice. A planner with ingenuity will find that esthetic needs can also be satisfied at little or no extra cost. With the materials available today, there is no need for a laboratory to look drab or austere.

FLOORING MATERIALS

An accomplished architect once recommended a certain type of rubber tile for a laboratory floor because his data indicated its superior resistance to acids. He had not bothered to find out whether or not acid spills would be a problem in this case. They would not, in fact. Solvent spills, on the other hand, were quite likely to occur, and the recommended tile had poor solvent resistance. This example illustrates two things: the importance of the laboratory operator's involvement with details that are sometimes overlooked even by experts, and the need to study each laboratory's requirements individually.

There is no flooring available with perfect resistance to chemicals, so the choice will in all cases be the result of a compromise. Flooring manufacturers' data on chemical resistance should be studied, keeping in mind how the materials offered will stand up against the types of chemicals which will be most commonly used in *this* laboratory.

Vinyl

Vinyl flooring, either in sheet or tile form, has shown good performance in laboratories where it has been installed for several years. Vinyl exhibits good resistance to most inorganic chemicals and to many aliphatic hydrocarbons. More vigorous solvents, however, especially ketones, will attack or soften it. Strong oxidizing agents can cause rapid discoloration, and heavily colored organic compounds will often penetrate into the vinyl flooring and cause permanent stains. Damage in all cases will be reduced, of course, if the spill is removed at once.

When vinyl is chosen for laboratory floor use, the heavier commercial grade is recommended over the thinner material often used in homes. The cushioned type, while easier to walk on, does not have sufficient ability to withstand laboratory wear and tear. Neither does the no-wax type. Good maintenance should include regular waxing, which will greatly increase the ability of any flooring to resist damage from wear and from chemical spills.

Sheet vinyl is best installed by professionals, whereas tiles can be laid by amateurs if instructions are followed carefully. Sheet flooring is usually considered more attractive and easier to maintain. It has few seams to open up in time and become dirt catchers or entrance points for moisture. In some materials, seams can be "welded" by heat, thus forming a permanent seal. Others will in time require additional sealer. Tiles, on the other hand, can be laid with less waste and also have the advantage of being replaceable if damaged. Whichever type is chosen, a supply of extra tiles or a few scraps of sheet flooring should be kept for possible future repairs.

Concrete

In one new laboratory, a concrete floor was given a very smooth finish and then treated with a sealer until it lost all porosity. Several types of concrete sealers are available, some with high chemical resistance. Such a floor is easy to maintain and to repair if damaged by strong chemicals. Rubber or vinyl floor mats, available from hardware stores and other supply outlets in roll form, can be placed in front of work benches and in other traffic areas. They are comfortable to stand or walk on and have good chemical resistance. Routine maintenance consists of mopping and an occasional hosing down outdoors. Such mats might also be considered for use on other types of flooring. They will catch most spills and are easily replaced when worn out or damaged.

Rubber Composition

Rubber is vulnerable to attack by many organics and should be avoided by laboratories using certain solvents. On the other hand, it has a high resistance to many inorganic chemicals and is often preferred in facilities where such materials are likely to be spilled. Like vinyl, it is available in both sheet and tile form.

Other Materials

The other usual flooring materials (wood, ceramic tile, etc.) are rarely seen in laboratories. Planners should be aware, however, of new developments in this field and investigate them with the same question in mind: How will this material resist the chemicals to be used in this particular laboratory?

Different types of flooring may be used in different parts of a laboratory. In a large university chemistry building, one material was used for the organic laboratories and another for the inorganic because of the different chemicals handled.

A word should be added here about laboratories where highly

flammable vapors are present. Conductive flooring can be installed, as it is in hospital operating rooms, to avoid a build-up of static electricity and subsequent sparks.

WALL TREATMENT

Concrete, cement block, and wallboard are the most common wall materials found in laboratories. All can be painted after treatment with an approprite sealer or filler. The choice of paint is very important. Both durability and ease of cleaning must be carefully evaluated. "Bargain" paints are rarely a bargain in the long run, particularly when labor is such a large part of the cost.

Paint Quality

Painting is generally performed by a contractor on a bid basis. Unless another agreement is made, a contractor will furnish the lowest priced paint that will do an acceptable job, even though it may have inferior durability. To avoid this, the contract should be very specific with regard to the type of paint to be used. It may even specify a certain brand.

In the laboratory, alkyd paints are preferable due to their durability and cleanability. Although latex paints have come a long way since their development, they do not measure up to the alkyd variety in laboratory applications. The popular latex enamels, in particular, are inferior to a good alkyd enamel. They are hard to apply smoothly and they pick up dirt far too fast.

Semi-gloss (also called satin finish or eggshell) alkyd paints are the most suitable for laboratory wall application. They have neither the easy dirt pick-up of flat paints nor the shine of high gloss enamel. Although flat paints are better able to hide irregularities in the wall surface, this advantage is not enough to recommend them. High gloss paints are best in laboratories where clean-room conditions must be maintained or where frequent cleaning has to be carried out.

Prior to painting, wallboard is sometimes given a heavily textured surface. This is impractical in a laboratory since it makes cleaning more difficult. The smoother stiple finish is attractive and easier to keep clean.

High Resistance Paints

Paints made for high chemical resistance are also available. One type is similar to regular alkyd enamel but has been modified for greater resistance. It has proved excellent for many laboratory applications where regular paints would not stand up, but even this type has its limitations.

When the highest resistance to corrosion and solvent fumes is needed, catalyzed paints are the answer. These come in two parts: a clear or colored base finish and a catalyst. Since their pot life is limited (typically eight hours), they must be mixed just prior to use. Brochures with information on their chemical resistance can be obtained from companies that sell industrial finishes. Since the resistance to different chemicals varies from brand to brand, a planner should study several types to find one that will best suit the particular application. These finishes are of the high gloss type. For highest chemical resistance to fumes, a coat of clear finish should be applied on top of the colored one. While the vehicle in the finish is very resistant, the pigments may not be and therefore could discolor.

In one laboratory, an operation that occasionally produced highly corrosive fumes was confined to an alcove with special ventilation. Some items were treated with a catalyzed finish. After five years, the finish showed no signs of attack, unlike other items in the same area that were finished in a more conventional way.

Catalyzed paints, while quite expensive, do have superior coverage as well as excellent protective qualities. They can be used not only on walls but on work benches, equipment, or any other paintable surface.

CEILING TREATMENT

In a testing laboratory with a room about 15x30 feet, the ceiling was painted just like the walls. The result was an acoustical disaster. With the hard floor, hard walls, and solid work benches, conversation across the room was quite uncomfortable. Over the phone, a person speaking from this room sounded as if he were standing in the middle of a cathedral. Another laboratory room with about the same dimensions had an acoustical ceiling. The difference was staggering. In fact, the changes that took place in the room's acoustics from hour to hour during the ceiling installation were dramatic.

Even a small room will benefit from acoustical ceiling treatment. A veterinarian's examining room was only about 7x7 feet, but with its hard walls and ceiling and no upholstered furniture to break up the sound waves, conversation was uncomfortable. Proper ceiling treatment would have made a great deal of difference.

There are many types of acoustical tile to choose from. It is easily applied by a do-it-yourselfer to an existing ceiling. The tiles must be handled with great care as they are fragile.

In many cases, a suspended ceiling is installed in a laboratory, since the existing ceiling of an industrial type building may be higher than needed. The lower ceiling brings about significant savings in both heating and cooling costs. In addition, it provides space for bringing in utilities overhead. The suspended ceiling consists of T-shaped aluminum extrusions holding boards of acoustical material, usually 2x4 feet in size. Flush mounted fluorescent light fixtures may be made part of such a ceiling. The ceiling boards are readily removed for access to utilities.

Acoustical ceilings do have certain drawbacks. They are fragile, porous, and attractive to dust. Areas surrounding air intakes will soon show dust deposits, but periodic vacuuming with a soft brush attachment will keep a ceiling in good shape for a long time.

The dust-catching properties of regular acoustical materials make them unsuitable in places where dust may cause problems, such as where clean-room conditions must be maintained or where microbiological work is performed. For such applications, special ceiling boards covered with plastic film are available. The plastic vibrates with the sound and transmits it to the interior, where it is absorbed. The plastic surface is easy to clean by conventional means. A sealing tape is available to prevent dust from entering along the edges of the boards. The cost of these ceiling boards is considerably higher than that of the regular ones.

LIGHTING

The installation of appropriate lighting should be considered part of the ceiling treatment. Planning for this is best left up to a specialist, who will see to it that work areas receive the proper amount of light from the right directions.

Fluorescent lights are now used almost universally. They may be recessed in the ceiling, as they usually are when suspended ceilings are used. Surface-mounted lights are used on hard ceilings and are considered more efficient.

Bare fluorescent light tubes are likely to cause uncomfortable glare. This has often been overcome by the installation of translucent plastic sheets beneath the lamps. Unfortunately, the sheets may absorb a significant amount of light. In more modern fixtures, the light is usually dispersed by sheets of molded clear plastic containing small prisms, which eliminate glare while absorbing very little light.

Fluorescent light fixtures vary in quality. Lower priced ones often develop an annoying hum after being in operation for a time, and the transformers may have a limited life span. High quality fixtures are more expensive but are a good investment.

Fluorescent tubes come in different colors. "Cool white" is most commonly used in laboratories. "Warm white" is recommended where fluorescent and incandescent lights are used in the same room. "Daylight" tubes give off light that is not felt to be comfortable by most workers. The line spectrum of the light emitted from tubes will distort colors, so another type of light should be used in areas where color judgments are to be made.

INTERIOR DECORATION FOR THE LABORATORY

A typical laboratory is often a rather dreary place to work, which may have a serious effect on worker morale. There is no reason why this should be. The cost of creating a more pleasant environment is very reasonable, but it does take some planning and effort.

Color Choices

Colors in a laboratory should be coordinated, just as in a home. If pre-finished work benches are to be installed, they might set the color scheme. While they are available in several colors or combination of colors, the choice is not unlimited. In one case, the laboratory operator was color blind, so his wife took over the job as decorator. First, she selected a two-color scheme for the work benches. Color chips in hand, she then chose a floor covering from a number of samples submitted. For the wall paint, she found a standard color of the recommended quality that harmonized with the cabinets. A few appropriate charts and a colorful cloth wall-hanging of pipes and valves completed the decor. The result received many favorable comments from visitors to the facility.

The use of light colors on large surfaces improves room light markedly. A light-colored floor is no harder to maintain than a dark one, but the difference in appearance is great.

When the walls of a large laboratory were scheduled to be repainted, the laboratory personnel decided to do it themselves provided they could choose the colors. The department secretary was appointed color coordinator. She came up with a daring but artistic scheme. Over a weekend, employees moved in with rollers and brushes, and by Monday morning the laboratory had taken on a new look. An added bonus was that very little equipment had to be moved off the benches. Knowing which items were fragile or sensitive, they were able to work around them.

In the corner of another laboratory, there was an unsightly array of pipes and conduit. This was also a dust catcher. It was easily covered up with plywood panels, which were painted to match the surrounding walls.

In another case, a wooden structure was built to hold equipment. A paint dealer matched the work bench color, which made the painted structure look like an expensive built-in.

Some laboratories may even want to consider using wallpaper in a limited area. An appropriate pattern of solid vinyl wall covering, which is both resistant and very washable, can be a cheerful addition when used in the right place. Wood paneling in an office area is another possibility that should not be overlooked. Little extras like these can make a great deal of difference in comfort and morale, usually at a modest cost.

7

Work Benches and Fume Hoods

Work benches and fume hoods constitute a major cost factor in any laboratory. A careful study of available products and good planning, however, can bring about significant savings with no loss in serviceability.

SUPPLIERS

Distributors of general laboratory equipment also offer one or more lines of work benches and fume hoods, often shown in a separate catalog. In addition, there are several companies specializing in this field. The annual LabGuide issue of *Analytical Chemistry*, published by the American Chemical Society, has a good listing of suppliers. A laboratory planner should obtain catalogs from several sources and compare both features and prices.

The supplier should be able to take care of installation or to recommend a local contractor who can do it. Very simple installations are sometimes performed on an in-house basis.

CABINET CONSTRUCTION

Cabinets come in a variety of units to suit individual needs. These are then bolted together to make benches. Any combination of drawers, cupboards, and knee-hole units is possible. Some fume hoods come complete with their own base cabinets. Where desirable, desk-height units can be integrated into the system. All units come with legs which can be adjusted for an uneven floor. Standard installation leaves room behind the counters or along the middle of peninsulas for utilities. After the cabinets are installed, counter tops are put in place and bench-mounted fume hoods added.

Wood versus Steel

Laboratory cabinets are available in both wood and steel. Educational laboratories often use wood, while industrial laboratories usually prefer steel. Wood cabinets do not have the sterile look of steel, but they are far less resistant to physical abuse. They come prefinished with a wood stain. A damaged area is not difficult to refinish. Wood cabinets have no rust problems, but some users have complained about poor chemical resistance of the interiors. The quiet operation of doors and drawers with no metallic noises is appreciated by many.

Some wood furniture is now faced with plastic laminate. Its resistance to chemicals is superior to that of most regular finishes, and color choices are wide. Best of all, such surfaces are very easy to keep clean. In case of damage, however, repairs can be difficult.

Steel cabinets have superior resistance to physical abuse. Refinishing scratched surfaces can often be done with a spray can. Finishes used on steel also have a high resistance to chemicals, even on the interior parts, but long-term storage of reagents that give off corrosive vapors will eventually lead to rusting.

Gone are the days when steel cabinets were available only in gray or olive drab. Some laboratory planners have created interesting color schemes by using one color for cabinets and another for doors and drawer fronts.

Sound-deadening material is used to make door and drawer operation quieter than formerly. One manufacturer features drawer fronts of molded plastic, which makes it possible to have the handles recessed.

At the University of California in Berkeley, the building manager of Latimer Hall selected wood for benches where inorganic work was performed and steel for organic work. This decision was based on his experience with finishes at the time the laboratories were installed.

In the long run, the choice between wood and steel is usually a matter of taste. Prices are quite comparable. When asked why he had chosen wood benches for the recently built chemistry laboratory of the U.S. Geological Survey facility in Menlo Park, California, the supervisor simply replied, "Because I like wood!"

Assessing Quality

It is hard for a layman to judge the quality of the various cabinets offered. Much of the quality is hidden beneath the surface and will not show up until the cabinets have been in use for an extended period of time. Price is only a partial measure of quality. One reason for this is that high style cabinets, offering no more quality or durability than the regular types, demand higher prices.

When comparing different cabinets, there are a few clues as to which ones provide the best value. First of all, a manufacturer of laboratory furniture should have data on the chemical resistance of finishes used. Even though nobody expects to splash reagents over the fronts of the cabinets, accidents do happen. Resistance data should be compared to a list of chemicals to be handled in the laboratory. A costly finish with outstanding resistance may not be needed in all cases.

Sturdy door hinges and drawer suspensions are a necessity, and there is a definite difference between manufacturers here. If a manufacturer has a display of cabinets, the prospective customer should act like a buyer of a used car, slamming doors and yanking open drawers. He should then make a careful examination of the various components and compare them with what he has seen elsewhere.

Most useful of all is to see examples of the furniture under consideration after it has been in use for some time in another laboratory. A sales representative may be able to arrange for this. This is once again a time for door slamming and asking questions. How well does the finish stand up? What are the comments from those using the furniture? Do doors and drawers stay shut after closing?

Kitchen Cabinets

Many small laboratories use kitchen cabinets, also available in wood or steel, and are quite happy with them. Normal prices of high quality kitchen cabinets are not much lower than those of the laboratory variety, but special bargains are often available. They come in far fewer types of units, which make the laboratory benches less adaptable to specific uses. Finishes are not made to resist chemicals, and the overall construction is seldom of the heavy-duty type. In a microbiological laboratory where they have been used for years, however, workers have been pleased with them. They were adequate for the light duty service and were purchased when a local dealer had a clearance sale.

Unfinished wood kitchen cabinets can sometimes be found at attractive prices. Before making a price comparison, the planner should determine the cost of finishing, which could be considerable. The unfinished cabinets usually come without hardware, which can add a great deal to the price if purchased separately.

WORK BENCH COMPONENTS

The general layout of work benches, including hoods, was described in Chapter 2. Now comes the decision about where there should be drawers, cupboards, or other features. Small, shallow drawers, for instance, are just right for storing hydrometers and similar pieces of equipment and should be close to where these are to be used. Burets also need shallow drawers, but they must be wide. Certain pieces of glassware need fairly deep drawers. Much equipment calls for cupboards of certain sizes.

Catalog in hand, the laboratory planner places units containing drawers and cupboards where they should go. The plan should be marked off with the catalog numbers and will later be given to the cabinet supplier. Some suppliers offer planning kits to make the job easier.

A work bench rarely consists of a solid bank of cabinets. So-called knee-holes are placed in appropriate locations for sit-down work. Stools can be stored in them when not in use. They are also good for keeping waste containers out of the traffic pattern.

In one laboratory, an atomic absorbtion instrument was placed on top of a three-foot knee-hole. Drainage from the instrument's atomizer went into a bottle underneath. Data from this instrument was recorded on a strip chart recorder placed on a typewriter stand. When in use, the recorder was at the operator's right, where it was easy to observe and adjust. When not in use, it was kept out of the way in the knee-hole, along with the operator's stool.

In another laboratory, a vacuum pump was installed on a shelf built into a two-foot knee-hole well above the floor level. It was out of the way, off the floor, and easy to service. Noise was substantially reduced by attaching rubber stoppers to the pump's base plate as vibration absorbers. Connection to equipment was through a hole in the work top.

These are just a few examples of creative uses for knee-holes.

They can be put to many other uses, depending on a laboratory's specific needs. Unfortunately, they too often end up as storage areas for items that should be stored elsewhere.

WORK TOP MATERIALS

There are many different types of work tops available for laboratories with prices varying over a wide range. The important point is to select a material that will stand up under the conditions in a given laboratory, since replacement is very costly and repairs often next to impossible.

At one time, tops were made from solid hardwood and impregnated with a material that would seal the surfaces and provide good chemical resistance. Many recipes were given for such treatment. The 1959 edition of CRC's *Handbook of Chemistry and Physics* showed recipes containing ingredients such as coal tar pitch, phenol, and "good fresh aniline oil." In all fairness, it must be said that those tops stood up quite well. Their heat resistance was poor, but burn marks did not show too readily on the black surface. In addition, they could be sanded and refinished in case of damage. They were also gentle on glassware.

Stone Tops

Tops made from slabs of Alberene stone have enjoyed a great deal of use despite their high cost. Their temperature resistance is excellent, and so is their resistance to chemicals. Their main disadvantage is porosity. Spilled liquids are readily absorbed and the results are quite unsightly. Various top dressings have been recommended to reduce this problem, but they must be applied at regular intervals.

Cement Composition

Cement composition tops have about the same properties as natural stone but at a lower cost. They have a more even appearance, lacking the veins found in stone. In these, too, porosity may be a problem. Treatment with vinyl and similar plastics works well against porosity and many chemicals but is not effective against stronger solvents. Pressure treatment with epoxy compounds, on the other hand, has produced tops with outstanding properties. Though costly, such tops have become very popular.

Solid Epoxy

Tops made from solid epoxy compounds are resistant to just about any kind of chemical abuse but are very expensive. They are often sold with an integral backsplash and curved junction, which makes cleaning easy. They are much easier on glassware than either stone or cement composition.

Plastic Laminate

Extensively used in kitchens, plastic laminates have found wide application in the laboratory. Their main advantage is comparatively low cost. Their main disadvantage is poor heat resistance, coupled with the fact that repairs are difficult to make. On the other hand, their chemical resistance is surprisingly high. Heavily colored organics, though, must be quickly removed. If allowed to penetrate into the material, they may cause permanent stains. Some manufacturers feature a special laboratory grade plastic laminate for which detailed information on chemical and heat resistance is published. Some of these are not much higher priced than the kitchen variety. Most laminates today have a satin finish, as opposed to the high gloss used in the past. This minimizes the visibility of surface scratches at the cost of slightly poorer cleanability.

Compressed Wood Fiber

Countertops of compressed wood fiber have the advantage of low cost. Although their resistance to chemicals and some liquid spills is very poor, they could be considered for areas where no liquids are handled.

Ceramic Sheets

Ceramic in sheet form is a spin-off from the space program, where materials capable of resisting extreme conditions are required. Its cost is high, but so is its resistance to both heat and chemicals. Laboratories often employ it in areas where other materials will not stand up. This material is similar to that used in solid kitchen cook tops.

Ceramic Tile

In European laboratories, ceramic tile is very popular, but it is rarely seen here. The cost is well below that of monolithic slabs but considerably higher than that of plastic laminate. Its resistance to both chemicals and heat is outstanding. In case of damage, single tiles may be replaced. Modern materials used for grouting are also very resistant to chemicals. The slightly uneven surface does not seem to cause any complaints among users.

Stainless Steel

Easy cleaning and sanitizing make stainless steel popular in food or biological laboratories. Contact with strong mineral acids, however, will show that it is not quite "stainless." An abrasive is required to remove such stains. While resistant, a steel top will temporarily bulge when exposed to heat, which sometimes can interfere with the alignment of equipment set-ups.

CHOICE FACTORS

Some laboratory planners insist on using the same type of work tops throughout. Like the Mikado, a prudent planner will"let the punishment fit the crime" and thereby cut costs substantially. In one laboratory, the center work table, where heavy-duty work was carried out, was covered with an expensive monolithic slab. The rest of the tops were covered with much less expensive plastic laminate. Of course, work rules had to be established as to what types of work could be performed where, but this was a minor inconvenience. Years later, the tops showed no signs of damage.

In another facility, bench tops in the main room were covered with epoxy-treated monolithic material to resist the hard wear anticipated. The side room, where instrument work was to be carried out, was equipped with tops of plastic laminate. This resulted in substantial savings. Somehow it was difficult to convince the engineering firm in charge of building the laboratory that monolithic material, which they had recommended throughout, was not necessary everywhere.

In a case where a fume hood was to be placed on a portion of counter faced with plastic laminate, the planner found an economical solution. Since the plastic laminate was not suitable for use with a fume hood, he ordered a sheet of monolithic material ¼ inch thick and had it cut to the exact dimensions of the hood. The cost was just a fraction of that of a full thickness top. The sheet was put in place and the edges were treated with a silicone compound. This treatment stood up against highly corrosive materials as well as heat for many years.

Delivery schedules must also be considered. Monolithic sheets are generally cut to exact size at the factory. Depending on the laboratory's location, delivery may be slow or occasionally go wrong altogether. Half of the tops destined for a new laboratory in California went to Alaska, while the Alaska tops ended up in California. This caused serious delays in two building projects.

The other types of tops are fabricated locally, although the specified material may not be available locally. There are also tops sold in standard lengths and cut to size on location. It takes considerable skill to make smooth joints in such tops.

WORK TOP PROTECTION

Contact with even quite mild chemicals over a long period of time will damage most counter tops. Drips running down the sides of reagent bottles have left rings on many laboratory shelves, and drips from burets containing 1N acids and alkalies have eventually etched ceramic tile. A prudent laboratory operator can prolong the life of any work top, however, by taking a few precautions.

Inexpensive plastic dishes can be placed under reagent bottles. There should be beakers beneath burets when they are not in use. Some laboratories cover their reagent shelves with thin sheets of ribbed polyethylene, which is available in rolls at a reasonable price.

For the ultimate protection from chemicals, certain work areas may be covered with thin sheets of self-adhering Teflon, available in roll form. One version is sandwiched with a thin sheet of aluminum foil, which increases heat resistance by conducting heat away from high temperature spots.

FUME HOODS

Before selecting a fume hood (appropriately called "stink hood" in old chemistry books), a laboratory planner must study performance requirements. What types of fumes are to be removed? Smoke, unpleasant odors, water vapor, flammables, corrosives, toxics, or strong oxidizing agents? The hood must be designed to handle the most hazardous or corrosive materials anticipated, even though these will be present only rarely. Catalogs give good information about the capabilities of the various types of hoods. Federal, state, or local legislation may also dictate hood requirements for certain kinds of hazardous work. A company selling hoods should have information available on this but will require detailed information about the work to be performed from the laboratory operator.

Hood Construction

Traditionally, hood interiors were made from asbestos sheets, which had both good chemical and good heat resistance. Their main problem was that they absorbed any liquid that was spilled on them. Asbestos has now lost its favor because of its toxic properties. Modern hoods usually have interiors of highly resistant plastic, often fiberglass reinforced. Such surfaces are easy to keep clean, but their temperature resistance may be limited.

Most hoods come with sash-type doors that move vertically, exposing the whole opening when they are open. Some have doors that move sideways, which allows them to be only half open at any time. Traditionally, doors have been made of shatter-proof glass, but heavy clear plastic is now common. Although this has the advantage of high impact resistance and light weight, its temperature resistance is inferior. Also, the plastic is subject to attack by some organic vapors.

Hoods usually have a light built into a sealed opening, making it explosion-proof for use with flammables. Fans used for flammables must also be explosion-proof. Those used for corrosive fumes must be made of material resistant to such fumes. The ductwork must also be corrosion resistant, a matter often overlooked. Built-in exhaust fans make installation and servicing easy, but some tend to be noisy. Remotely installed fans are generally quieter and may serve more than one hood. The chief problem with fans is that they are often improperly serviced. Lubrication and belt tightness must be checked as specified by the manufacturer, or costly repairs will result.

Installation

Installing a hood with its own fan against an outside wall rather than in the roof could result in considerable savings. The exhaust must then go through the wall in a place where it will not be objectionable in the area outside the building. Not all odors are unpleasant, of course, as workers in an industrial laboratory next to a potato chip factory discovered. Every time the wind blew from the south, they were treated to tantalizing aromas. They threatened to demand free handouts if the situation was not corrected.

Even when a hood has an integral fan with guaranteed performance, the installer will check the face velocity. For some types of work, laws may require minimum velocities. A highly visible draft gauge should be installed on a hood to inform personnel about its operation. This is especially important where remote or very quiet fans are used.

Building a Simple Hood

A laboratory can sometimes save money by building a simple hood where no hazardous, flammable, or corrosive fumes are involved. In one case, a hood was needed solely for drawing off unpleasant fumes from a muffle furnace and water vapor. A carpenter constructed a plywood box which reached from counter top to ceiling and had a large opening in the front. This was painted with a resistant paint, primarily to seal the surfaces and make them easy to clean. An inexpensive fan exhausted the hood to the outside. This simple hood worked for many years.

There is one problem with that type of hood, however. Some day a worker will want to use it for work it should not handle, such as evaporation of flammables. The laboratory operator must set strict rules as to its use. "Insignificant" amounts of materials that could be potentially hazardous may in time become substantial amounts when new types of work are added.

Ductless Hoods

Ductless hoods are equipped with built-in fans and filter systems which trap the volatiles being removed. In this manner, the spent air can be returned to the laboratory, thereby eliminating the need for exhaust ducts. The filter system must be tailored to the types of fumes to be removed. Its use is therefore limited and must be strictly controlled. Since it is independent of a permanently mounted duct system, a ductless hood can be moved to a new location in the laboratory if required. Its mobility also makes it well suited for use with modular laboratory furniture.

MODULAR FURNITURE

Several manufacturers now offer modular furniture. Each module is complete with base cabinet, work top, and utilities, such as sinks and electrical outlets. The modules can be placed in banks as required and then hooked up to rough plumbing and wiring. With modular furniture, it is easy and inexpensive to change a laboratory arrangement as requirements change. They are most often seen in research laboratories for that reason. Their main disadvantage is their high cost.

A less expensive variation of modular furniture is in use in some European laboratories. The bench frames are built of permanently installed steel channels complete with tops. Modules containing combinations of drawers and cupboards can be hooked into the system and changed as needed, but utility outlets are fixed.

8

Utility Outlets

Bringing in the various utilities, such as water, gas, and electric power, has been discussed in Chapter 3. While the work benches are being installed, the utilities will be hooked up to suitable outlets, or inlets in case of the sewer. This is another area where good planning can produce significant savings with no loss in performance or durability.

MOUNTING OUTLETS

Utilities may be mounted behind work benches and brought through the tops where needed, or they may be mounted above the benches. The first method conceals wires and pipes, which gives a neat and uncluttered appearance. Its disadvantage is that repairs and modifications may be difficult. In one laboratory building at a large state university, some of the base cabinets did not have removable backs, so a hole had to be cut in order to repair a water leak. The building manager now insists on having all utility outlets mounted above the work benches whenever any of the laboratories are being remodeled. Even though most laboratory benches do have removable backs, work on utilities mounted behind them still takes some crawling and handling of tools in cramped spaces.

This problem can be avoided by mounting utility outlets above the work benches. This method also has the advantage of keeping the benches clear of obstructions and easy to clean. Of course, exposed water pipes and electrical conduit do not have much esthetic value, and they are dust catchers as well. A coat of paint, however, both improves their appearance and makes cleaning easier. An even better solution is to box them in and cover them with removable panels.

In an educational laboratory such overhead mounting must be planned with care. The typical work benches here are either peninsulas or islands where the utility box would go down the center with a shelf on top. The total height must be such that the instructor is able to look over the shelf in order to supervise a class. To complicate matters, there may be a waste water trough along the center of the bench which requires some free space above it. These multiple requirements present a challenge to the designer.

HOT AND COLD WATER

Laboratory supply houses have faucets of high quality and durability that are able to stand up under heavy duty conditions. They require little service and their chrome plating will take considerable punishment. The main disadvantage is their high cost. If the faucet is mounted on the countertop, a goose-neck tube will bring the outlet high above the sink to facilitate rinsing of tall glassware. Good quality kitchen faucets have done well in many laboratories but are often too low when mounted on a countertop.

Mixing faucets are very convenient, though some laboratory workers prefer separate faucets for hot and cold water. Many like the aerators commonly used in kitchens, while others prefer faucets equipped with rubber or plastic hose.

One laboratory operator installed faucets with long handles for easy control, the type commonly seen in hospitals. These proved very popular with the staff.

Regular washer-type faucets are hard to control for low water flow, such as required by condensers. For these purposes, needle valves are recommended. They give excellent control, and flow rate does not change as long as the pressure remains constant. The inexpensive plug-type valves are rarely adequate for laboratory water service.

The local plumbing code may require that laboratory faucets be of the anti-siphoning type, which prevent water from being drawn back into the system in case the pressure should fail. This type is standard in home washing machines and dishwashers.

One laboratory used industrial-type brass valves for many of their water outlets, rather than the more expensive chrome plated ones available from the laboratory supply house. While their appearance was different, their performance was the same. To make the valves look better to both laboratory workers and visitors, the manager would occasionally treat them with metal polish.

As any home owner knows, no water faucet or valve is maintenance free. A laboratory operator should for that reason have repair parts and proper tools available at all times.

DEIONIZED WATER

As mentioned in Chapter 3, DI water has now replaced distilled water for most laboratory purposes. PVC pipe, which is both inexpensive and easy to install, is normally employed to carry the water to points of usage. It is very important that the lines do not have "dead legs" where water is allowed to stand, since this could encourage bacterial growth. In cases where sterile water is needed, special steps must be taken. The water may be boiled, which also drives out dissolved gases, or ultraviolet radiation may be employed, a common method in the cosmetics industry. Membrane filters will also effectively remove bacteria. Laboratories needing sterile water often use two bacteria removal systems of different types mounted in series. Any such system must be mounted as closely as possible to the point of water usage.

Laboratory supply houses feature sink-mounted faucets heavily plated with inert metals for DI water. Less expensive plastic valves, available from plumbing supply houses, may be used as long as they contain no metal parts. Self-closing valves, while more expensive, will substantially reduce DI water consumption.

Another way to reduce consumption is to install a pressure reducer ahead of the DI water tanks. This will also reduce wear on DI water valves.

It should be noted that plastic pipes and valves used for DI water have limited structural strength and must be well supported by straps and brackets.

SINKS AND DRAINS

Manufacturers of laboratory equipment feature special sinks and drains with high chemical resistance. Sinks made from slabs of Alberene stone, which were popular for many years, often had traps made of lead. The softness of lead made the traps hard to clean out, and the job required tender loving care with a wrench of just the right size.

Plastics have now taken over. Polyethylene sinks are quite popular, although some users have reported their tendancy to crack. The more expensive solid epoxy sinks have performed very well. In one laboratory doing cryogenic work, however, workers felt their resistance to temperature shock could be better.

Sink traps of resistant plastic are now well accepted. They are easy to install and also easy to clean. Drains of glass pipe are sometimes seen in special installations. Some laboratories install regular thin-walled sink traps designed for kitchen use. The author's experience with them, even in the home, has been poor.

Heavy-gauge brass traps are also available, sometimes on special order. They deliver quite good service. For most uses, however, plastic drain systems seem to work best.

The need for good chemical resistance is no longer as great as it used to be. Under current waste water rules, most powerful chemicals may no longer be disposed of in a sink. Good resistance, however, will be needed in cases where sinks do not drain to the sewer but to a waste holding tank which is pumped out by a waste disposal service.

With this exception, regular enamel sinks, preferably of the cast iron type, are satisfactory for most laboratory use. They should be mounted in such a manner that they may be removed and replaced in case of physical damage. Sooner or later somebody will drop a heavy object into the sink and crack the enamel. These sinks are very resistant to staining and easy to keep clean.

Stainless steel sinks, which are often seen in laboratories, can take considerable physical abuse without damage and, as mentioned earlier, are easier on glassware than most other materials. Even small amounts of some chemicals, however, such as dilute mineral acids, will stain them. The stains can be removed with an abrasive cleaner.

Laboratory sinks should be deeper than those commonly used in kitchens in order to make glassware rinsing easier and to minimize splashing. A trip to a nearby plumbing supply house will reveal a large variety of sinks available.

Flush-mounted cup sinks are commonly used to receive small to moderate amounts of water. These are often made of polyethylene. They can be made from 3-inch PVC pipe cemented in flush with the countertop. A reducing adapter will bring the diameter down to the standard drain pipe size.

Drain troughs are often put along the center of the island or peninsula type work benches, sometimes with water outlets directly above. This arrangement is good in educational laboratories because it keeps the students from having to leave their work stations each time they need a sink.

Traditionally, sewer drain pipes have been made of cast iron. It has a surprisingly high resistance to corrosive chemicals and has stood up well for many years, even before the days of strict waste water regulations. Where permissible under local codes, plastic drain pipes are popular today. Their price is reasonable and so are installation costs. Their resistance to inorganic chemicals is very high, but even small amounts of some organics may be absorbed and cause weakening, leakage, or other problems.

GAS

Inexpensive plug-type valves may be use for hook-ups to appliances that have their own gas flow adjustment. For those that do not, needle valves are recommended for precise control.

COMPRESSED AIR

Compressed air is often required in the laboratory. In some cases, it comes from a compressor beneath the work bench, while in others, the compressor may be at some remote location in the building.

Compressed air invariably contains water and usually oil from the compressor. It is also likely to pick up dirt and rust from the distribution line. For use in the laboratory, the air must be clean, particularly if it is to be used in instruments. One laboratory operator suggested the use of corrosion resistant pipes for the air distribution system in a building. This was turned down because of its high cost. The result was a heavy load of rust which began to accumulate in air filters after only a few months of operation.

Another laboratory cleaned the air in two stages. First, it went through an inexpensive commercial filter with easily replaceable cartridges. This took care of the bulk of the contaminants. It was then sent through special filters specified by equipment manufacturers, which were expensive but seldom needed changing because of the previous filtering.

One laboratory had surprisingly few problems with its instrument air, which was supplied from a commercial compressor with no special equipment. This laboratory was on the second floor of a building, while the compressor was at ground level. The air was taken upstairs through a pipe much larger than would ordinarily be used for such service. Since the air velocity was therefore very low, the pipe acted like a rectifier column and sent most of the contaminants back to the compressor air tank. As a result, the instrument filter needed very infrequent servicing.

In a less fortunate situation, air for the laboratory instruments came from an overhead distribution system, often with high air velocities which carried along contaminants. Servicing of this air cleaning system was both frequent and costly.

A compressed air line generally terminates with a valve and a pressure regulator. Such regulators are available from laboratory supply houses, but less expensive ones from stores selling welding supplies may often be acceptable.

ELECTRICAL OUTLETS

Electrical outlets can be installed in one of several ways. Small pedestals holding two or more outlets are often placed along the rear of wall-mounted work benches. They may also be put along the center of peninsulas or islands. Wiring is from below the countertop, which makes modifications difficult, just as with plumbing. While easy to reach, pedestals clutter the top and interfere with cleaning.

Some benches have an integral low shelf to the rear along the wall. The solid front of such a shelf is an excellent location for outlets. Wiring can be serviced or modified by removing the top shelf.

Some laboratories have installed outlets on the front of work benches, just below the top. Dangling wires from such outlets are at best a nuisance and could also be hazardous. This method of installation is not recommended.

Normally, outlets are not placed inside fume hoods, where they may suffer from corrosive effects or present an explosion hazard. Many hoods will permit outlets to be mounted on either or both sides of the exterior, somewhat above bench level.

Surface-mounted outlet strips may be mounted on the wall above a work bench or along the center shelf of an island or peninsula. With this system, one box for rough wiring will serve a number of outlets. Outlets are at regular intervals, such as 12 or 18 inches. In one case where such outlet strips were installed along two walls of a room from the same box, two circuits were put into it, feeding alternate outlets. This made it possible to supply power-hungry equipment from adjacent outlets. It should be noted that the finish used on some of these strips may have a poor resistance to even mildly corrosive fumes, so a protective finish might be advisable.

COMPRESSED GASES

Hydrogen, oxygen, nitrous oxide, acetylene, or other compressed gases are often used for laboratory instruments. If a gas is used in only one location, the cylinder is usually brought to that spot. However, there are many cases where a gas is required in more than one room. It will then be more practical to distribute it from one central location, preferably close to a building entrance. It must be where the cylinders will not block an emergency exit in case of an accident. One large laboratory had a separate room with direct access to the outside for gas cylinders. A ventilation fan kept air circulating in the room at all times.

Local building codes should be consulted with regard to safe storage and use of compressed gases. Pressure regulators must be compatible with the type of gas being handled, and distribution lines must be made of materials that will not corrode or otherwise be damaged by the gases they carry. Lines must be carefully pressure tested before being placed in service. If at any time one type of gas is to replace another in a line, the line must first be throughly bled. It must also be ascertained that the new gas is compatible with the line materials.

At the end of each work day, one person in the laboratory should have the responsibility for turning off the main cylinder valves for all gases.

CONCLUSION

The laboratory operator is offered many choices in selecting utility outlets. A careful study will need to be made in order to select those that give the best combination of performance, durability, safety, and cost for service in a particular laboratory.

9

Final Plans and Construction

Time has now come to transform the preceding plans into reality. All ideas and suggestions have been turned into drawings and specifications from which the laboratory can be built. Or have they? This is up to the laboratory operator to determine by carefully checking over all details so that corrections can be made before construction starts. Since the typical laboratory operator will usually not be familiar with many of the architectural terms and symbols, he should ask questions whenever something is not quite clear. No detail must be taken for granted. The case of the suspended ceiling will illustrate this:

In order to provide easy access to overhead utilities for repairs or modifications, the laboratory operator had specified a suspended ceiling, which to him meant one with large pop-out panels. On the plans, the designer definitely indicated a suspended ceiling, but the matter was never discussed in detail. When the ceiling was installed, the laboratory operator was out of town on other business. Upon his return, he found a solid ceiling anchored to a steel grid, which, though suspended, allowed no access. Five months later, it had to be torn into in order to modify the ventilation system. Such misunderstandings are not uncommon, even among professionals of the highest caliber.

CHECKING FINAL PLANS

When the architect or designer presents the final drawings, the main layout and all details must be checked against earlier plans and instructions. All dimensions must be in agreement with those specified. Slight discrepancies in door and window locations, for instance, could have serious consequences. Any item which can be identified by catalog number, brand name, etc., should be so indicated on the drawings or in the written specifications.

There are two words to watch out for: *or equivalent.* What is called equivalent by a designer or contractor may not be equivalent to the laboratory operator. Sometimes this problem arises when a specified item has been superseded by a "new and improved model." However new or improved the new model may be, there could be some reason why it would not be right for a particular application. Any substitution must be viewed as suspect. It is always wise, of course, to listen carefully to all suggestions from building professionals and evaluate the reasons they give for a proposed change, but nothing should go unquestioned.

Even the most reliable contractor or subcontractor may not be familiar with the requirements of a laboratory. He bases his bid on standard materials and procedures that he is accustomed to using for general construction. On the other hand, he may also throw in sophisticated and costly extras that are not required. It is for these reasons that everything must be well defined and cross-checked. Specifying materials and components by trade name is recommended. In many cases, of course, other brands could be acceptable, but it should be agreed that a switch can not be made without the laboratory operator's knowledge and approval.

Two examples will show how carefully plans must be studied. In one case, a stipple finish was requested for the laboratory walls. Although this had been put in writing, the memo slipped through the cracks along the way and never reached the drawings. The walls ended up with a heavily textured finish, the painting contractor's standard treatment for commercial walls.

The other example concerns the installation of separate faucets for hot and cold water when mixing faucets had been requested. The laboratory operator would have caught this had he checked faucet model numbers on the drawings.

Sometimes the designer or architect goes overboard on extras. In one laboratory where the fire hazard was low, the operator had asked for doors with small windows between rooms. Instead, solid UL-rated doors were installed, even though they were not required by local building codes. Worse yet, the doors were self-closing, a feature neither required nor desirable in this situation.

What all this means is that the laboratory operator must examine all plans and specifications with the greatest care and ask a multitude of questions, no matter how minor they may be or how stupid he feels they may sound. It has been said that the only stupid questions are the ones that are never asked. This would certainly apply here.

CONSTRUCTION

Construction will be supervised by the architect, designer, engineer, or general contractor. This does not mean that the laboratory operator can now relax or take a vacation. On the contrary, he must be available at all times, ready to handle the crisis of the day and send out distress signals whenever something does not seem to be right. To some, this may seem as if he is interfering with the work of building professionals, but he is really not. After all, he is the person who is going to live with the laboratory when it is finished and be criticized for anything that does not turn out as it should.

Misunderstandings

It must be pointed out that problems are rarely caused by people not doing their jobs right, but rather by misinterpretation and lack of communication. Many unexpected details will come up as construction progresses which require consultation among the concerned parties.

Take the case of the laboratory planner who specified that a shallow hood should be placed above a bank of drying ovens to carry away fumes. What he had in mind was a simple sheet metal box extending about two feet down from the ceiling. The designer, in turn, specified a much costlier pyramid-shaped affair, which was his concept of a hood. Fortunately, the sheet metal contractor questioned the design as he was getting ready to build the hood. After discussion with the laboratory operator, he built a simpler one at significant savings.

In the above case, the laboratory operator had given the designer oral instructions for the hood but did not see the sketch the designer made later for the sheet metal contractor. If the laboratory operator had put his instructions in writing or made a rough sketch, there would have been no misunderstandings.

Expecting the Unexpected

There are sometimes cases where space requirements, in spite of the most careful planning, just do not come out the way they look on paper. This can often be the result of last minute revisions. It may be discovered during construction, for instance, that it is impossible to fit utility connections into a tight space allotted to them. Any changes required must be scrutinized by the laboratory operator, since they may have to be made in a certain way in order to accommodate future operations.

Ceiling construction in one laboratory was such that the contractor wanted to move light fixtures slightly for more economical wiring. Before giving his approval, the laboratory operator insisted on having the lighting designer check the new plan for possible interference with light distribution in the room.

Many surprises surface when a laboratory is being installed in an existing building or when it is being remodeled. Even when existing facilities have been taken into account, some things do not show up until construction is under way. A concealed pipe, for example, which is shown on the original drawing of the building, may not be there. There might also be good news, such as in one laboratory where the sewer pipe was in the "wrong" place, which made it possible to connect it to two sinks instead of one. It even had a capped-off T just where it was needed. Such a discovery could not have been made until after work actually had started. Sometimes unused electrical conduit has been found inside walls, resulting in savings on installation cost. On the negative side, things such as leaking water pipes and heavily corroded sewer lines have also been found on occasion.

Safeguarding Equipment Information

Some contractors do not leave behind the information packed with equipment they install. After all, they probably think, what use will the laboratory people have for installation instructions? This material, often separate from use and maintenance instructions, will be very helpful down the line when repairs or modifications have to be made. The laboratory operator should collect all such information and file it away with care. He may never look at it again, but there could be a day when it will be urgently needed.

An installer may even inadvertently toss away warranty information that comes with a piece of equipment. In one such case, an air conditioner broke down less than one year after installation. It took phone calls to the designer, the general contractor, and the sub-contractor to get matters straightened out. Meanwhile, the July weather was getting hotter by the day.

In conclusion, it may be said that the laboratory operator must be on deck at all times during construction. No matter how good the architect, designer, or contractor may be, when construction begins, Murphy's Law will surely prevail: Anything that can go wrong, will.

10

Equipment and Supplies

The laboratory is built, work benches are in place, and utilities are hooked up. All is now in readiness to receive the equipment and supplies to begin operation.

EQUIPMENT SOURCES

Laboratory equipment and supplies are available from several sources, such as laboratory supply houses, manufacturers' representatives, mail order houses, and local retail stores.

Laboratory Supply Houses

A supply house, which generally represents several manufacturers, may be nationwide, regional, or local. Many offer a broad range of supplies, while others specialize in certain fields, such as medicine or education. A supply house usually has a "will call" desk, where an urgently needed item may be picked up within hours after the order is placed.

Factory Direct

Much equipment is now sold directly by the manufacturer

through a network of representatives. This is particularly true for high priced instruments. Such items may also be offered on a lease basis.

Mail Order Companies

Mail order companies often offer excellent service. Their catalogs sometimes list specialty items not sold elsewhere, such as unique pH electrodes or plastic products. Both laboratory supply houses and mail order companies may sell popular types of instruments under their own private labels at substantial savings.

Science Shops

Small retail stores called "science shops" are found in many larger cities. While their prices may be on the high side, they offer over-the-counter service for many items needed in a hurry. They will also sell glassware by the piece rather than by the case, a service often desirable for small laboratories.

Hardware and Electronic Stores

Local suppliers of hardware and electronic parts often offer many items useful in the laboratory at very competitive prices.

Drugstores and Supermarkets

At large drugstore chains and supermarkets, the operator of a laboratory can find great bargains on items such as paper towels and cleaning products when they are on special sale.

WHERE TO OBTAIN INFORMATION ON SUPPLIERS

Scientific and technical journals carry advertising by laboratory supply companies. Many have reader service cards that can be mailed in for further information. Since service may be slow, some of the advertisers have toll-free phone numbers for quick response to questions. Local supply houses are listed in the yellow pages in the phone directory. All it takes is one call, and the sales representative will show up, catalogs in hand. As already mentioned, the annual *LabGuide*, published by the American Chemical Society, is also an excellent source of information on suppliers.

SELECTING EQUIPMENT

Selecting equipment for a new laboratory calls for a careful study of catalogs and many discussions with sales representatives. As a rule, each item considered will have its good and its less desirable points. Choosing the most suitable make and model for a specific laboratory will often involve a compromise.

If a sales representative offers to arrange for a visit to a laboratory where equipment of the type being considered is in use, such an offer should not be turned down. A discussion with an actual user will be very fruitful. Some manufacturers show their equipment in display rooms, giving the prospective buyer a chance for true "hands-on" experience.

Electronic developments have revolutionized many types of laboratory equipment. Few instruments with dials are seen today, for instance. Digital readouts have taken over. While easier to use, estimating the amount of fluctuation of unstable signals is more difficult on a digital readout. There are also cases where an instrument's readout is more sensitive than its detection device, thereby causing unstable readings. For certain applications, a dial instrument may still be preferable. Electronic balances are now the rule rather than the exception. They are not only easy to use, but they generally require less service than mechanical ones. Prices have been considerably reduced in recent years.

Power supplies to instruments have come a long way since the old days of the electron tube. Solid state devices develop very little heat, which was a problem in older equipment. However, many solid state devices are sensitive to sudden voltage fluctuations, or they may suffer damage if the line power fails and suddenly comes on again. Much of the newer equipment may be connected to data processing systems, very important where a high volume of work is performed.

Selecting the right equipment has become quite difficult today because of the tremendous variety available. In addition, the planner must look into the future. Will a certain piece of equipment be suitable for future work? Here is a typical example: A forward-looking laboratory operator insisted on purchasing a pH meter that also could be used for specific ion determinations, even though such work was not being done at the moment. It took some time to get management approval for this more expensive instrument, but within 6 months specific ion determinations were started. A study of catalogs will show new trends in equipment design, which will soon make older types obsolete, even though they have not yet been discontinued.

SELECTING SUPPLIES

A new laboratory will need a start-up supply of a variety of consumable items. How much should be purchased? While it may be tempting to buy large amounts of items such as glassware and reagents while managment is still in a spending mood, it may also be unwise. First of all, there is storage to consider as well as the fact that many reagents have limited shelf lives. On the other hand, prices at start-up could be very favorable, since supplies will come as part of a general bid. Such pros and cons must be carefully weighed. Note also that management may have definite limits on inventory of supplies.

COST ESTIMATES

Starting up a new laboratory means preparing a long and detailed shopping list. Everything, down to the smallest spatula and hose connector, must be listed. This takes time and much attention to detail. Again, the check list of laboratory operations in Chapter 1 comes in handy. For each operation, everything that is needed should be listed down to the size and type of filter paper. Then the list is consolidated, since many items are used for more than one kind of work. Sometimes buying a different size or model of an item will make it do double duty.

What will it cost? A laboratory operator should be very careful about discussing figures with management until estimates have been made. Prices given in catalogs may not be current and should always be verified. Possible discounts from regular prices should not be taken into account unless they are guaranteed. Some instruments may require certain accessories that are sold separately. One laboratory, for instance, almost overlooked ordering the ultrasonic cleaner needed for periodic servicing of an instrument component. In another case, special glassware was needed for the operation of an instrument. The cost of minor items, such as reagents, glassware, and thermometers, adds up faster than most planners realize. The idea is to present management with a realistic start-up budget. Adding items after a budget is approved will involve a good deal of persuasion.

ORDERING

The manner in which orders are placed varies from one laboratory to another. In a large organization, the laboratory operator will prepare a requisition, which goes to the purchasing department, often via a department head for approval. Copies of the requisitions should be kept in the laboratory, which should also receive copies of actual purchase orders. In a small laboratory the operator may do the ordering himself.

Obtaining bids from two or more suppliers is highly recommended for larger orders at start-up. The prices quoted may come as a pleasant surprise, since a supply house will put its best foot forward when bidding on equipment for a new laboratory. If they get the business, they will have a foot in the door when it comes to orders for future day-to-day supplies, which is their bread and butter. Bids should be gone over in detail, not only for correctness, but also for possible substitution of some items.

A laboratory supply house will require credit information before accepting an order from a new customer. This is generally a simple procedure.

RECEIVING EQUIPMENT AND SUPPLIES

One laboratory operator said that it felt like Christmas when equipment and supplies for his new facility began to arrive. It was like a reward for all those hours spent selecting the right items and justifying their purchase. But receiving large amounts of equipment all at one time is a great deal more work than opening presents.

With each shipment comes a packing list with reference to a purchase order. The laboratory copy of the purchase order, a copy of the requisition, and the master list of all equipment and supplies must be on hand. The packing list is usually a copy of the acknowledgement of the order. All the items on an order may not be shipped together. Some may be back-ordered, while others may be shipped from a different location.

The first step is to examine the exteriors of all cartons and packages for signs of damage in transit. Then, everything is opened and examined, including items that will not be used for some time. If there should be any claims for damage in transit, these are generally sent to the freight forwarder, not the supplier, within the time limit for filing complaints. Fortunately, such damage is rare.

Every item is checked off on the packing list, which will become an authorization for payment, and also on the master list. Usually, the packing list is sent to the purchasing department after an authorized person has signed it.

The purchase price of some instruments may include set-up and check-out by a manufacturer's representative. In such cases, the laboratory operator should only check for physical condition of a shipment and call the representative about its arrival. Other equipment will generally be set up at its permanent location as soon as it is unpacked and checked. Supplies will be taken to their designated storage areas. The accompanying warranty cards should be filled out and mailed as soon as possible.

Equipment instructions must be properly and promptly filed. A separate file should be set up for each piece of equipment and should include the smallest note packed with it. A lack of such files is an invitation to future problems. In one instance, a strip chart recorder that had not been used for several years seemed just right for a procedure but needed some minor repairs and parts. After a couple of hours of diligent search, the instruction book was finally found tucked away in a drawer. It could easily have been missed. In other less fortunate cases, it took time-consuming correspondence with the manufacturers to produce information that should have been immediately available in the file cabinet.

For some instruments, the manual may be needed with the instrument for proper operation. This is particularly true in educational laboratories. In such cases, the manufacturer will generally be happy to supply extra copies, or the significant parts of a manual may be copied on the office copy machine. In all cases, the original manual and instructions must be properly filed.

CAPITAL EQUIPMENT

As equipment and supplies arrive, the laboratory operator will be contacted by the accountant, who will want to know which items are to be capitalized and how they are to be amortized. This is very important for tax purposes. Rapid amortization will usually offer many advantages, but IRS will question write-offs that seem to be too quick.

Capital equipment is durable and not consumed during operation.

In time, of course, it will suffer wear and tear and eventually become unusable. Its life will also be shortened by obsolescence, a very important factor. A 50 year old two-pan analytical balance, for example, may be in perfect working order, but its value is only that of an antique. The cost of equipment will also determine whether or not it should be capitalized. Rules for this will vary from one laboratory to another.

It is up to the laboratory operator to estimate the *effective* life of a piece of equipment. He must be able to justify such an estimate in case of a challenge from IRS. This calls for detailed knowledge of the equipment and its use, plus some idea of how soon it may become obsolete.

EQUIPMENT IDENTIFICATION

All laboratory equipment of any appreciable value should carry some type of identification. This should be arranged for as soon as the equipment arrives. It is particularly important for capital equipment in order to keep accounting records straight. Permanently attached metal tags with numbers are often used but may be hard to attach to some types of laboratory equipment. Smaller items of relatively low value could also carry some type of identification. They are easily marked with an engraving tool available from hardware stores. Many police departments will loan engraving tools at no charge as part of their theft prevention program.

Laboratory equipment is sometimes stolen. Most popular are smaller items of relatively high value, such as electronic balances. Permanent identification marks definitely discourage theft. One stolen microscope was quickly returned to its owner when it appeared on the used equipment market. It was easily identifiable because its owner had engraved marks not only on the body, but also on objectives and eye pieces. The thief, fortunately, had ignored them.

11

The Laboratory in Operation

Construction has been completed and equipment is coming in waiting to be set up, but the most important elements of the laboratory are still missing. Nothing can be done until people are present to run the equipment, interpret the tests, and write the reports. At this point, some time before the laboratory is scheduled to begin operation, important personnel decisions must be made.

STAFFING THE LABORATORY

A laboratory is only as good as the people who work there. Choosing the right workers and giving them proper training is of the utmost importance. Qualified personnel may be found in several ways:

By advertising in the newspaper classified section.

By listing the job openings with an employment agency.

By placing a notice in professional publications.

By notifying college placement services.

By word of mouth.

A newspaper ad, particularly in a paper with wide circulation, will no doubt reach the most people. In this or the other written job announcements, the employer should always give name and address along with a brief job description. Sometimes, for one reason or another, a company chooses not to reveal its identity. Such blind ads, however, usually have the effect of discouraging the best qualified applicants. This practice is frowned upon by the American Chemical Society and most newspapers. Listing a phone number, where one can call for an application or further information about the job, will help screen out most unqualified applicants.

If the position is listed with an employment agency, the agency personnel should be given as many details about the job qualifications as possible and be instructed not to send people for interviews who do not meet the requirements.

For professional level positions, a professional journal is a good place to advertise nationally. A notice can also be placed with a local or regional division of the society.

Universities and colleges operate placement services for their graduates, usually at no charge. College science students also make excellent temporary or part-time laboratory workers while still going to school. They are eager to get experience in the field which will help them get a permanent job when they graduate.

Some of the best job openings never reach a newspaper or employment agency. These are the ones spread by word of mouth from one laboratory to another or by people who have heard of the new laboratory about to open.

Application Forms

Larger companies have their own application forms, carefully edited to meet all legal requirements. An applicant must not, for instance, have to reveal age or ethnic origin. As everyone is aware, of course, a form which asks for an applicant's date of graduation from high school is just as revealing of age.

Smaller organizations may want to use standard application forms from stationery stores, although these often have too little space for all important information.

Resume or Letter

An applicant with some experience may prefer to supplement the form with a resume. Even an inexperienced applicant should be required to write a letter accompanying the application form. The way in which such a letter is written will give a prospective employer a better idea about the applicant. It should be noted that reasonably good writing ability is needed for almost any job. Professionally written resumes are easily spotted by a person going through a stack of applications. What would such a person have written by himself or herself?

The Interview

An interview must be conducted privately in a relaxed atmosphere. It takes skill to conduct a good interview. A typical applicant may be quite nervous. In a larger organization it is often performed by the personnel department in the presence of the potential supervisor. At the end of the interview, the applicant should have a chance to ask questions about the job and the employer.

A promising applicant should be given a tour of the facility. During such a tour, he or she will be introduced to current workers, whose comments should not be overlooked.

Details about salary, benefits, and vacation policy must be carefully explained. In a larger company, these may be discussed in a brochure. Some companies offer profit sharing plans and retirement plans, which may be attractive to many applicants.

Final Selection

After reading applications and conducting many interviews, the laboratory operator faces the crucial decision. He checks references, but they may only shed partial light on an applicant. After all, one will only give references that will be favorable. Checking with past employers may bring about interesting information, but it may be misleading. There was the case when an employee was fired for drinking. Feeling sorry for him, the employer gave him good recommendations.

Credit references could be very informative when evaluating a prospective employee.

Final selection is the result of weighing many factors. First of all, the qualifications must be right. Will the applicant fit into the atmosphere of the laboratory? Is there evidence showing that he or she will get along with the existing staff? Finally, does the applicant want a permanent position or just a stop-gap proposition?

Notifying Applicants

The successful applicant will, of course, be notified by mail or phone about the decision. All the others must also be told as soon as possible. A simple form letter is all it takes. An applicant goes through much work preparing an application and deserves the courtesy of a reply. Far too many employers, often smaller ones, ignore this. Often such a letter includes the sentence: "Your application will be kept on file in case future openings should materialize." Such encouraging words should not be used unless the employer really means them. If an employer indeed does keep an active file of applications, the letter should clearly state the length of time such applications will be held.

Job Responsibilities

Each laboratory worker should be clearly instructed as to duties and responsibilities. A written job description outlining such matters in detail may be very desirable. However, it must be amended in writing should any changes come up. Certain menial tasks which must be performed should be evenly distributed.

Staggered working hours may often be of advantage, provided that this does not cause undue hardship on workers. If an employee prefers to come in early and leave early, such a request should be taken advantage of, as it will extend the laboratory's effective working day.

On the Job Training

There is often work in a laboratory for which no previous training or experience is needed. For these entry level jobs, the laboratory operator will be the instructor. Even people with experience in other laboratories, though, will need some training in the specific aspects of a new job. Sometimes poor work habits will have to be corrected. More details on training will be found in the chapters on laboratory records and safety.

REPORTS

A typical laboratory will produce many types of reports. If a computer is available, these will come forth almost automatically. If not, they will have to be typed. In a typical case, an analysis report, for instance, will be written up by hand, taken to the office for typing, and returned for checking and signature. Typing a report in the laboratory will save time and effort but, unfortunately, many professionals consider it beneath their dignity to operate a typewriter, though not a computer, which uses essentially the same techniques. Longer reports will usually be typed by a secretary, often from hand written notes.

ORDERING SUPPLIES

Ordering initial equipment and supplies has already been described. Policies for re-ordering of supplies consumed in daily operations vary greatly. In a large company, a requisition goes to the purchasing department, which does the actual ordering, often after the laboratory operator has discussed the matter with the supplier by phone. A top limit is usually set for orders that can be placed without special approval. Many companies also have a lower limit for purchase orders since the office work required is the same regardless of order size.

Under more informal conditions, the laboratory operator will obtain a purchase order number, call the supplier to place the order, and then send the information to the office.

Many laboratories have a "want book" in which workers list items that are in short supply before the shortage becomes critical.

SUPPLY SOURCES

For day-to-day supplies, a good relationship must be established with a supply house, preferably a local one. If one company does not handle all the items needed, more than one will have to be used. A long-term relationship with supply houses offers many advantages. First of all, significant discounts may be offered. An outside salesman will drop by periodically and become familiar with the laboratory's operation. Based on such information, he may be helpful with recommendations.

An alert salesman will keep tabs on what a laboratory buys, noting, for instance, frequent repeat orders for a certain reagent. In such cases, he might suggest ordering in large quantities less often at a lower price. However, it should be noted that management may be sensitive to having a large inventory of supplies on hand.

A good salesman will also keep his customers informed of new products related to specific types of work performed. He should not present items that are not needed just because they are new.

The author had a pleasant experience with a distributor he had worked with for many years. Authorization had been given for the purchase of a new muffle furnace. Calling the inside salesman, it was established that the selected model was in stock, ready for immediate delivery. Then the salesman said: "But I won't sell it to you!" The reason? Scanning his computer, he discovered that they had one left of an older model, which they were willing to part with at a much lower price. It was perfectly adequate for the intended purpose. The laboratory operator was praised by management for the savings, but it was really the effort of a salesman who wanted to maintain a long-standing relationship that made the bargain possible.

DEAD FILES AND OLD SAMPLES

File drawers become overcrowded much too fast. Periodic clean-out of active files is a necessity, but very time consuming. Inexpensive cardboard file boxes are good for permanent storage of dead files.

Sample shelves also become overcrowded quickly, but it is often required to keep samples for extended periods of time. Older samples can be stacked into cardboard cartons, clearly identified by laboratory project references or quality control numbers, and stored at a convenient location away from the laboratory. Periodically they should be inspected and the out of date ones eliminated. One word of caution, however. Under current pollution control rules, laboratory samples may not be indiscriminatly disposed of as garbage. The laboratory operator should call in a waste disposal service when in doubt.

LABORATORY HOUSEKEEPING

A clean and orderly laboratory improves performance and productivity. It also enhances morale and has a beneficial effect on safety. Unfortunately, housekeeping in many laboratories leaves much to be desired. Dirty glassware, for example, is often left on the countertop until someone has to clear it away because the space is needed or because no clean glassware is left.

Setting the Rules

The laboratory operator should establish the rules for housekeeping and set a good example for all who work there. When working on a project, he should put all reagents and equipment back into proper storage. It is a great morale builder when workers see that the boss cleans up his own mess. Besides, they will quickly come to his assistance should he be in a bind. Far too many laboratory supervisors will routinely walk away from the debris of an experiment or a test and expect others to clean it up.

A good rule in the laboratory is that everything must be cleaned up at the end of each day. Glassware should be washed and placed on a drying rack. Reagents should be put on the shelves where they belong. Equipment not needed the next morning should be placed in proper storage. Coming into a laboratory in the morning and finding the previous day's dirty glassware all over a work bench is as inspiring as walking into the kitchen the morning after a dinner party.

Janitorial Services

A large organization will have its own in-house janitors. For smaller facilities, janitorial services will usually be taken care of by a contractor. The exact work to be done by such a contractor should be spelled out in writing, where no details must be left out. Unfortunately, the degree of cleanliness is hard to define in exact terms.

Usually, a new janitorial service will start work over a weekend. On Monday, everything shines. In many cases, the quality of work will gradually taper off. One laboratory had to change services twice during one year for this reason. A potential contractor should give references, which must be checked out, and show evidence of liability insurance.

A janitor surrounded by sensitive instruments may be like a bull in a china shop. In one laboratory, the janitors were not allowed to clean anything above floor level, including windows, without supervision by laboratory personnel. Janitors should not be allowed to handle containers of flammable or hazardous waste. Laboratory personnel should clear the floors completely when the janitor announces that it is time for floor waxing.

Cleaning Materials

A distributor of laboratory supplies will offer a number of cleaning products. For most jobs, however, home-type cleaners are satisfactory and lower priced. In some cases, though, an instrument's warranty may be voided if the cleaning materials recommended by the manufacturer are not used.

Abrasive cleaners are often needed for laboratory jobs, such as removing pencil markings from etched spots on glassware. No glassware should be put away with such markings. Liquid cleaners of this type may not be as effective as powders, but their gentler abrasives are less likely to cause scratching.

Self-adhering labels are often used for temporary identification. They are easy to peel off but may leave a sticky residue. This is easily removed with isopropyl alcohol.

Cleaning brushes of all sizes and shapes are found in catalogs from laboratory supply houses. A generous supply of the types needed should be kept on hand.

KEEPING TRACK OF COST

What does it cost to run the laboratory? The laboratory operator will be required to know this. He will also probably be required to submit annual budget requests and be expected to stay within the budget. On the other hand, he will also have to answer many questions should his estimates sound too high to those in charge.

Some costs are easy to estimate, such as wages and salaries or utilities. The cost of equipment service is often overlooked, as is the cost of replacement parts. Lamps for atomic absorption instruments, for example, have finite lives and are quite costly to replace. Electrodes for pH meters and other instrument components all need periodic replacement.

The cost of reagents should remain essentially constant if the laboratory keeps performing the same type of work. Unavoidable glassware breakage should not change too much from year to year. Prices of all these items have risen substantially over the years, so allowance for price increases must be made. Anyone who has been ordering silver nitrate for some period of time has seen an extreme example of this.

In educational laboratories, the consumption of supplies per student should be quite predictable for any given course. The cost of equipment maintenance and replacement, however, is often underestimated. Unfortunately, in a college budget system, the reward for saving money one year may be a reduced budget the following year.

In a research laboratory, expenses are quiet unpredictable. One never knows just where a project will go and what may be needed. When a project budget is required, making it up may be a formidable task which calls for both extensive knowledge in the field and diplomatic ability.

Establishing a budget is simpler in a testing laboratory. When the cost of a certain analysis is calculated, however, some items are often overlooked. Of course, it will include man-hours needed. If an instrument is used, the cost of its amortization and maintenance must be figured. If the analysis is non-standard, there is the cost of developing the proper method. The cost of reagents and supplies is easier to establish. If a client requests that a report be prepared in a special way, this will be an extra cost item.

A manufacturing company approached the cost of quality control in a realistic manner. Before a quotation was made for a new product, the control chemist submitted an estimate of the time required for running the necessary tests. An hourly charge was established for laboratory operation, and the cost of quality control could thus be established. The sales department did not like the arrangement because the cost was very much the same whether they sold 200 or 2000 gallons of a product.

It is up to the laboratory operator to make such realistic cost estimates. If he does not, he will sooner or later find himself in a serious bind. The main challenge is to make practical estimates of those hidden costs and distribute them in the proper manner.

LABORATORY ETHICS

Anyone working in a laboratory will have access to information that is not public. If the employer is engaged in development work, for instance, news of such work should be confidential. A new employee may be required to sign a secrecy agreement under which all developments made by the employee become the company's exclusive property. When signing such an agreement, the author made certain that some formulation principles which he had developed prior to employment with that company were excluded.

In an analytical laboratory, there must be complete confidence between laboratory and client. All results belong to the client and must not be disclosed to others. There was a case, for instance, where some special methods were developed in connection with a nutritional study. This involved an analysis of monkey feces. The laboratory did not announce that it had newly developed methods, since these really belonged to the client.

Medical laboratories, of course, must never disclose any results except to the patient or the doctor in charge of a case.

An analytical laboratory receiving samples of products under development by a client will often become aware of what the client is working on. Secrecy is mandatory in such cases. In fact, having a secrecy agreement will usually be preferable. Both client and laboratory personnel will then be able to discuss the project freely among themselves.

There is always the possibility that there may come a time when a laboratory discovers itself doing work in connection with illegal activities, such as drug manufacturing. Here is where the whistle must be blown. If the case is not reported to authorities, the laboratory could become a co-defendant in a prosecution.

As a rule, analytical laboratories will not reveal the names of clients unless authorized to do so. Notebooks and samples should not be kept where visitors are likely to see them.

12

Laboratory Records

A few years ago, a customer filed suit against a nationally known manufacturer of automotive brake fluid for supplying faulty merchandise. When the company's control chemist went before the jury with his detailed quality control records, he was able to prove that the allegedly faulty batch made two years earlier met all the requirements established for this product. The plaintiff lost his case.

The importance of good laboratory records is hard to overestimate. Everything must be recorded in a permanent manner and preserved so that it can be retrieved months or years later when the information may be badly needed.

The increasing use of computers has made record keeping less of a chore than it once was. Even in laboratories with computerized records, however, there is still need for some written record keeping. The same kinds of records must be kept regardless of the method used.

LABORATORY NOTEBOOKS

The use of loose sheets of paper for even minor observations or calculations must not be permitted. In a typical laboratory, each worker should have a personal work book where all operations are recorded and all calculations shown. The books should be hard bound and have numbered pages. Tearing a page out of a laboratory notebook or making an entry illegible, no matter how useless it may seem, must be considered a cardinal sin. The size of the pages and the number of pages should be such that the notebook can be conveniently carried from place to place. Most users prefer lined pages, but others may want squares or even blank pages.

Each entry must be dated. This not only makes it easier to retrieve information in the future, but it makes it possible to reconstruct the chronology of work performed and correlate it with other activities. In an analytical laboratory where a customer alleged that a sodium thiosulfate reagent solution used in a titration had been improperly prepared, the date reference made it possible one year later to tie the sample in question to a specific batch of reagent. This careful record prevented a threatened lawsuit.

When filled to the last page, notebooks must be filed in a safe place for posterity. In some laboratories, they are even locked away.

METHODS OF RECORD KEEPING

The way records are kept varies from one laboratory to another, depending on the type of work performed.

Research and Development

Special care must be taken with notes regarding development work that could lead to a future patent. A patent attorney should be contacted about such matters and his recommendations followed to the letter. In general, all details must be recorded, including experiments that did not work out. There are many cases where exciting developments have come as a result of something going "wrong." However, there may be a considerable time lapse, or the worker who performed the experiments may no longer be around. Good records make it easy to resurrect the old work.

In one such case, the author was saddled with the job of reviving an old project because the person in charge of it no longer worked at the laboratory. However, he had left behind volumes of notes, complete with a sprinkling of profanity indicating where things had not worked out. With such information, the project was quickly completed.

Laboratories working on products that eventually will need approval by a government agency, such as FDA, would do well to consult the agency in question about its latest requirements. If the request is made in writing, this can be of great help in case of an argument later on.

All samples and specimens from development work must be accurately labeled and cross-referenced to the laboratory notebooks, in which materials used are identified not only by chemical names but also by manufacturers' trade names. The author had a frustrating experience while developing a cleaning product. The original work had been performed a few years earlier and then the project had been dropped. For some reason, the old results were not reproducible. Old samples and records were recovered from the archives and examined. A phone call to the manufacturer of the key ingredient uncovered the problem. A slight change had been made in this ingredient which actually created a purer material. Without the old impurity, however, the product did not perform. The problem was solved in hours instead of days or weeks by adding the missing impurity.

Analytical Laboratories

A typical analytical laboratory handles a large number of samples, usually for routine tests. Keeping track of these samples is of major importance. Reporting results for the wrong sample could result in anything from the loss of a good client to a liability suit. Recording each sample before any work can be started is essential.

Sample Recording. A good way of recording is to keep all incoming samples on a designated table until they can be properly marked. Materials requiring refrigeration or freezing can be put into special sections of the refrigerator or freezer where nothing else should be kept. A note left on the sample table will assure that the refrigerated or frozen sample will not be overlooked.

Each sample must be given an identification number, which will be used in all analytical records. Self-adhering labels or labeling tape are commonly used. The ink should be waterproof. Regular masking tape works well for this purpose and is inexpensive. Labels with pre-printed numbers may also be obtained. After labeling, samples are duly recorded in the main record book, which is a running account of all work performed in the laboratory, or they may be recorded in a computer. The record must completely identify the sample, its origin, and what tests are to be performed.

Records of Tests. Either of two methods may be used for recording routine test results. In the first, measurements and calculations are recorded in the individual worker's notebook and then transferred to the main record book. In the second method, a separate work book is kept for each routinely performed test. Whoever does the testing enters the results in the book and initials it. The pages should be lined with columns to suit each test's requirements.

By reviewing the main record book (referred to in one laboratory as the "Good Book") at least once a day, a laboratory manager can keep track of how work progresses.

Industrial Laboratories

The most common duties performed by an industrial laboratory are testing of raw materials and testing of finished products.

Many laboratories find it practical to keep separate records for these, which means that the sample numbering system for raw materials must be different from that used for finished products. Other laboratories lump everything together for the purpose of record keeping but have separate storage areas for the two types of samples.

Quality Control Records. Accurate quality control records are of utmost importance and may be required by law, as in the pharmaceutical industry. A few years ago, mandatory "Good Manufacturing Practices" were proposed for the cosmetic industry. Although the proposal was later dropped, many manufacturers go by the rules worked out at that time for laboratory testing and record keeping. The food industry also requires extensive testing and record keeping.

Product Sample Records. Many industrial laboratories prepare samples of proposed products for customers to test. These are often not well recorded or identified. There have been many cases where it was impossible later on to say which version of a proposed product had been submitted. Every sample of this kind should carry both a date and a reference number. For the sake of simplicity, the number can be the page number from the appropriate work book. With such a system, all pertinent information regarding a sample can be quickly retrieved when needed—during a telephone discussion, for example.

Poor Record Keeping. Unfortunately, many small industrial laboratories have inadequate records. This can have unpleasant consequences. In one case, the manufacturer of an alcohol solution was challenged by a customer about meeting specifications set forth in the contract. For his defense, he had not retained samples that could be traced to the disputed deliveries, and his only written information consisted of scribbled notes on loose sheets of paper from his desk drawer. He lost the contract. An independent investigation indicated that he had probably done his job right, but he had no way to prove it.

Not all small manufacturers keep poor laboratory records, however. One very small cosmetics maker keeps elaborate records on par with the best systems seen anywhere. He claims that it gives him peace of mind and also makes a favorable impression on his customers.

PURCHASING RECORDS

In a small laboratory, purchasing records will usually be kept in the general office files. In a larger organization, they will probably be kept in the purchasing department. In either case, there should be copies in the laboratory, where they can serve many purposes. Reference to an old purchase order, for instance, will often make re-ordering quicker and simpler. At budget time, records of past purchases enable a laboratory operator to estimate future supply needs, making allowance, of course, for changing prices.

PERSONNEL RECORDS

In a large organization, personnel records are kept in the personnel department files. In a small laboratory, they may be somewhere in the general files. Such information should be considered confidential, with steps taken to limit access.

MAINTENANCE RECORDS

Laboratory equipment has to be maintained on a regular basis. A record should be kept of the date and the type of maintenance performed. This should be placed in the file for the instrument involved. If the maintenance is done on an in-house basis, the report should be initialed by the person who performs it. If done by a factory representative, there will be a receipt which should be filed. More than once, good maintenance records have saved an analytical laboratory from litigation when a client claimed that its results were incorrect.

RECORDS OF PROCEDURES

Procedures used in a laboratory must be recorded for easy reference, down to the smallest detail. An analytical laboratory will often use standard procedures from official publications, but to these must be added references to specific types of equipment used. As a result, a laboratory needs its own procedure manual to use in day-to-day work. Every operation should be described in detail, with references made to official methods where applicable. It takes thought and skill to write such a manual in a manner that will make is useful to all workers. Safety measures, where needed, must be included. Simple operations, such as the preparation of reagent solutions, should be specified.

Many laboratories use a numbering system for their procedures for easy reference. Procedures should always be dated, since they will some day be superseded by new versions. At that time, the outdated ones should be carefully filed. There may be cases where reference will have to be made to them in the future. One person should be in charge of the methods books and make sure it is kept up to date.

A suitably sized ring binder with a sturdy cover is recommended, preferably covered with plastic. This procedure manual must be kept where all who use it will have easy access to it. More than one copy may be needed in a large laboratory. Just in case an accident should take place, an extra copy must be kept in a safe spot. For durability and protection from spills each page can be inserted in a vinyl sheet protector. No pages must ever be "borrowed" from the book even on a temporary basis.

PAPERWORK REDUCTION

It becomes evident from the above that even a small laboratory will have a fair amount of time-consuming paperwork to take care of. More than one researcher has complained about the fact that record keeping interfered with creative work. Fortunately, there are ways to simplify the process.

Analytical laboratories may have the greatest amount of paperwork. Samples come in, are recorded and analyzed, and reports must go out. Copies must be kept of all reports. Billing information or reports on time charges for in-house work must be generated.

In one company's quality control system, one piece of paper served many purposes. A copy of a purchase order went to the receiving department. When merchandise arrived, it would be physically checked and the copy would be signed by a receiving clerk. The material would then be placed in quarantine, and the copy would be forwarded to the laboratory as notification that something had arrived. The laboratory would perform the required sampling and testing and then enter results in its own records. The purchase order copy would be signed by an authorized person in the laboratory and then go to the warehouse, thus authorizing the removal of the material from quarantine for transfer to the regular storage area. Finally, the copy would go to the office for processing by accounting and inventory control. All this with one piece of paper!

The only problem with this system was in cases where partial deliveries were made. For such situations, special paperwork had to be generated, but it was rarely necessary.

In a manufacturing company, the laboratory was responsible for generating manufacturing instruction sheets for all batches to be made. The sheet would go first to inventory control, where it would be ascertained that all ingredients were on hand. If not, the sheet stopped there and was held until the situation was corrected. It then went to the manufacturing department, where it was used as a work sheet. Finally it was submitted to the laboratory with a sample of the finished product. After testing, the laboratory would sign the sheet and release the batch. Again, all on one piece of paper. Afterwards, it went back to inventory control for posting of withdrawals and to cost accounting. It finally ended up in the production manager's files.

In another industrial laboratory, the company required a separate analysis sheet for every sample tested. The control chemist of a branch plant saw his files increase in volume by the day. Finally, he devised a report form which would hold data for about twenty samples. Besides reducing paper volume, this system made it possible to scan a large number of samples for trends in analytical results by looking at a small number of sheets of paper.

In an analytical laboratory, two copies of each report went to the laboratory manager. On this, he would show the charges to be made for the work done. He alone was able to justify extra charges for some samples and reduced charged for others. One copy was retained for the laboratory files and the other was used for billing. The notations with regard to charges were easily blocked out should more copies of a report be needed.

COMPUTERIZED RECORDS

In laboratories today, computerized records are common, particularly where large scale analytical work is being performed. Many instruments are designed to work with computers but may require interfaces, which are sold separately. One laboratory doing a large amount of commercial analysis work is now feeding all instrument data into its computer, which does all the calculations, even though this meant replacing an expensive instrument that would not feed into a computer. All reports from this laboratory are printed out by the computer in the form requested by the individual clients. The manager, though, insists on keeping a human touch. Each report is personally signed by the authorized person with a blue pen to make the signature stand out from the computer print-out.

For many purposes, high speed dot-matrix print-out is satisfactory. In other cases, such a presentation may not be proper. Letter-quality printers are now available for computers at affordable prices. In addition, many newer typewriters may be hooked up to computers for high quality print-outs. These may be slower than the computer printers, but the cost of such a typewriter is considerably less than that of an individual typewriter plus a computer printer.

CONCLUSION

No attempt has been made in this chapter to report the many ways a computer may serve in laboratory record keeping. What has been shown here are some of the types of records required in various kinds of laboratories and suggestions for handling them. Above all, the message of this chapter is a call for unfailing accuracy in keeping laboratory records, regardless of the method used.

13

The Laboratory Handyman

Among the many hats worn by the laboratory operator is sometimes that of general handyman. This is often true in a small laboratory, where assigned repair and maintenance personnel are not available. Here the operator soon becomes an electrician, a plumber, a painter, and an equipment rebuilder. In a larger institution, union rules may prohibit a laboratory worker from touching even the simplest hand tool. In cases, though, where union technicians are unfamiliar with the peculiarities of the laboratory equipment, an agreement is sometimes made under which nonunion laboratory personnel may perform certain types of specialized repair work that would otherwise require an outside contractor. If properly done, such in-house work can save substantial amounts of money over a short period of time.

LIMITATIONS

A good handyman must be able to recognize his limitations and employ professional assistance for work he is not completely familiar with. In the long run, this will prove less expensive. Too many hazardous wire connections and improper plumbing repairs have been observed in laboratories. One OSHA inspection of a large testing laboratory uncovered a myriad of electrical problems, which were corrected only at considerable cost. In another case, an improperly installed sink drain caused flooding of the laboratory and a potentially hazardous situation.

VALUE OF LABORATORY HANDYMAN

On the positive side, many cases can be found where laboratory personnel have installed and maintained equipment at the highest professional level and have even developed useful devices not commercially available. Their intimate knowledge of the equipment paid off in superior results from an inventive approach.

Often there is hidden talent amont the laboratory staff. One worker, for instance, had electronic testing equipment at his home which he brought in and used as required at no charge. A young analyst who had done woodworking with her husband at home turned out to be an expert at handling this type of equipment during a laboratory remodeling job.

RELATED EXPERIENCE

There are no college courses on how to be a handyman. This is learned through experience, often in connection with a hobby, such as woodworking or electronics. Many times it is acquired through repair and maintenance tasks performed around the house. Such experience is directly transferable to the laboratory, although some repair practices that may be acceptable in the home are not permissible in the laboratory. This is particularly true in case of wiring. All laboratory equipment, for example, must be connected with a grounded wire and a three-prong plug. Most building codes require metal conduit for industrial wiring, whereas nonmetallic cables are considered satisfactory for most home use. The common, though unsafe, home practice of patching a damaged cord must never be resorted to in the laboratory. Likewise, many home-type plumbing repairs may not stand up under the heavy-duty requirements of the laboratory.

SOURCES OF INFORMATION

Where does one obtain information on adapting handyman expertise to the more rigid requirements of the laboratory? For the simpler jobs, the public library has many books of the "how-to" type. Such a reference could also be purchased, making a useful addition to the laboratory library. Local fire and building departments are more than willing to pass on information regarding their requirements, which vary from one community to another. Equipment manufacturers, too, often have special instructions pertaining to repairs and maintenance not found in the regular operator's manual.

TOOLS FOR THE LABORATORY

If laboratory personnel are expected to perform repairs and maintenance, the proper tools must be available. This is particularly true in an educational laboratory, where stockroom personnel have a multitude of repair and maintenance jobs to perform. Only the highest quality tools should be purchased, as they will cost less in the long run. A cheap screwdriver, for instance, will soon begin to wear and damage screws that are hard to replace. It may even slip and cause injury. An "almost right" wrench will damage nuts beyond use. A cheap wire cutter may leave a few strands of wire uncut, an annoying and possibly hazardous condition.

Sources of Tools

Where are quality tools available? Laboratory supply houses sell handy tool kits for normal repair and maintenance jobs. They may come in compartmented boxes or pouches that are easy to carry to where they are to be used. Such organization also makes it simple to spot a tool left behind on a job. Unfortunately, kits are often quite expensive and may contain some tools that will never be needed, or they may lack tools required in a certain laboratory. In such cases, tools are best obtained separately from hardware stores or electronic supply houses. Since prices for high quality tools can vary considerably, it pays to do some shopping. So-called "bargains" should be carefully scrutinized for quality.

Tool Storage

Tools have a habit of "walking away." This may not mean somebody has stolen them. They could have just been left at the point where they were last used. Replacing a lost or missing tool, even an inexpensive one, could mean costly down-time for an instrument. To avoid such problems, there must be some kind of system whereby it can be ascertained that all tools are returned to proper storage immediately after use. A formal check-out system is recommended in any situation where more than a very small number of workers will have access to tools. Borrowing tools for weekend use at home should be prohibited or at least carefully controlled.

Special Purpose Tools

Some instruments come with special tools for adjustment or routine maintenance. These should normally be kept with their instruments for ready access. Using them for other purposes in other locations should be discouraged. In an educational laboratory, such tools should be labeled and kept in the stockroom unless absolutely needed during routine work.

Tool Rental

If the laboratory lacks the right tool for a one-time highly specialized application, it would be impractical to buy it. There are tool rental firms which have just about anything that might be needed: heavy duty sanders, large capacity drills, power saws, electrical conduit benders—the list goes on and on. Best of all, the prices are quite reasonable. Such services should certainly be considered.

ROUTINE MAINTENANCE

Maintenance is a never-ending task. Equipment manufacturers are normally very specific about periodic care of the products they sell, but many maintenance tasks are similar to those performed around the home. Minor repairs done in time will often prevent the development of unsafe conditions, as well as keep equipment in good working shape.

Faucets

Even the best water faucet will eventually start to leak. Repairs are usually quite simple, requiring parts available at the nearest hardware store. Caution must be exercised in tool selection, however, since the improper wrench will damage chrome plating, while the wrong size screwdriver will ruin the screw holding the new washer in place.

Sink Traps

Disassembling a sink trap may require a special wrench. It also requires caution, as the trap might contain toxic or corrosive materials or small pieces of broken glass. More than one lost item, such as a magnetic stirring bar, has been found during this process. If even the slightest sign of corrosion or other physical damage is discovered, the trap must be replaced.

Wires

When damage to a wire is close to the end, the wire can simply be shortened. Otherwise, any damage to the insulation calls for replacement of the wire. The new wire must have the same or higher current capacity and an equivalent type insulation. Connecting a new wire to an instrument is an interesting project which requires careful attention to detail. No shortcuts are permissible.

Electric Plugs

When there is an electrical malfunctioning, a faulty plug is often the culprit. It should be replaced only by a plug able to stand up under industrial use. A clamp should be used to hold the plug to the wire. This will avoid strain on the connection in case the plug gets pulled out by the wire. If possible, wire ends should be coated with solder before being attached.

Light Bulbs

Replacing a light bulb is a simple task, but it should be noted that the new bulb must not be of higher wattage. When replacing fluorescent tubes, special care must be used, since serious injury could result from breakage. It is not advisable to mix tubes of different colors when replacement becomes necessary.

EQUIPMENT MAINTENANCE

No laboratory equipment is maintenance free. Some requires regular servicing by the factory representative or by an expert with special tools or skills. Much maintenance, though, can be performed on an in-house basis by carefully following the instructions given in the manufacturer's manual. This would include jobs such as cleaning, lubrication, adjustments, and replacement of fuses, lamps, or switches. The slightest malfunctioning of an electrical switch, for example, calls for immediate replacement. All drive belts, such as those found on vacuum pumps, must be regularly inspected for signs of failure and for proper tightness. The vacuum pumps themselves need periodic oil changes, and many motors require lubrication, some with a specific lubricant recommended by the manufacturer. In fume hood fans, the most common cause for failure is lack of lubrication. All of these are important tasks that can be performed by laboratory personnel

MAJOR PROJECTS

A tour of laboratories will reveal much ingenuity on the part of laboratory operators and workers. Sometimes there is home-built equipment constructed at a fraction of the cost of what is available commercially. Sometimes there is equipment not even manufactured commercially at any price. Unfortunately, such equipment does not always measure up to professional standards. Safety features may have been neglected, poor quality materials may have been used, or professional type finishing touches may be missing. No major in-house project should be undertaken before ascertaining that both talent and tools are available for obtaining results which will be completely professional in all respects.

Examples of Laboratory Handiwork

The following projects may serve as inspiration to those who want to improve their laboratories on a tight budget.

Equipment Rack. One laboratory needed a rack along the center of a free-standing work bench for supporting condensers and other equipment. After checking the prices of ½-inch aluminum rods from laboratory supply houses, the laboratory operator discovered such rods were available for less in a hardware store. These were easily cut to length with a hacksaw and the cut ends were made smooth with a few strokes of a file. The rods were then attached to the countertop and ceiling with suitable fasteners ordered from a laboratory supply catalog, along with some standard clamps for attaching a variety of equipment.

pH Meter Stand. An otherwise satisfactory pH meter had one fault: it could be read only when in a straight vertical or horizontal position. A small box was constructed to hold it at a 45-degree angle for easy adjustment and reading. The material used was a piece of plywood left over from a home project. The box was carefully sanded, spackled, and primed. For a professional look, it was then given a coat of hammertone paint, followed by a clear finish.

Incubator. When laboratory work required an incubator with an internal electrical outlet for a mixing machine, such a thing could not be found in the catalogs. Even if it had been available, the price would have been way outside the budget. Fortunately, the laboratory operator enjoyed woodworking as a hobby. He went to work building a double-walled cabinet with insulation between the walls. An appliance parts house furnished a heating element of the right size, a laboratory equipment supplier a thermostat, and an electronic parts supplier a quiet and efficient circulating fan. The rest, including a fuse holder and an on-off indicator light, came from a hardware store. Both exterior and interior surfaces of the cabinet were carefully finished with paints of good chemical resistance for a professional look. This was a lengthy project, and the builder's impatient wife wanted to know how soon the "monster" would get out of the house. It finally did go to the laboratory, where it performed well for many years, retaining the name "Monster" forever after.

Some Negative Examples

On the negative side, some poor examples of laboratory handiwork have been seen. In most of these cases, a little extra work on the part of the handyman could have elevated the results from just passable to excellent.

Careless Workmanship. One laboratory operator combined a variety of equipment for a distillation procedure and finally came up with something that met his needs. In other words, it worked —but he had a tangle of wires, water hoses, and blocks of scrap wood for support. With less than one day's additional work, he could have improved both performance and safety features.

Dangling Cords. In another case, a "temporary" light duty extension cord was used to hook up a new piece of equipment with an adapter to by-pass the three-wire grounded outlet. A year later, the equipment was still there and so was the cord, now hanging from a hook in the ceiling.

Poor Quality Materials. One laboratory built an extension to an existing work bench. The idea was both logical and practical. But why did they use the cheapest grade lumber available, full of knots and knot holes? For this type of project, the extra cost of good materials would have been minimal. The addition was left unfinished, making it look still worse. A little spackle, sandpaper, and paint would have given the project a finished look.

SALVAGING FIRE-DAMAGED EQUIPMENT

Fire damage may present the biggest challenge of all to a laboratory handyman. One night, part of an industrial building containing the laboratory burned. It was quickly declared a complete loss, but was it really? The laboratory manager and his assistant sifted through the debris. Yes, much could be salvaged, but not without a great deal of work. Everything that looked recoverable was placed in cartons and put in a warehouse. Then replacement costs were determined. These came as a shock; the answer was to rebuild as much as possible.

Repair Methods

All equipment that could stand such treatment was soaked in dilute sodium hydroxide, which removed both fire deposits and old finish. All wiring and switches were removed, regardless of apparent condition. One by one, the pieces of equipment that were deemed salvageable went to the laboratory operator's home for rebuilding. Metal parts were finished with a resistant paint, actually superior to that used originally. Solder joints damaged by heat were repaired and all cooling water connections carefully pressure tested. Brighter indicator lights than those originally furnished were used. Fuses of just the right size were installed to minimize damage in case of shorts. Duplicating the special wires and switches proved to be the most difficult and time-consuming part of the job, but even that was finally accomplished with some perseverance.

Drying Oven

A drying oven looked hopeless after the fire, at least on the outside. However, the insulation had protected the interior parts, including the thermostat. With the outer metal shell cleaned and painted and the insulation, wiring, switches, and door gasket replaced, the oven looked and performed like new again.

A Learning Experience

The money saved by restoring this equipment was not the only reward. An added bonus was that the laboratory operator had become throughly familiar with his equipment. Sources of future malfunctioning could be quickly located and repair would be inexpensive and fast. This would save both money and costly down-time in the future.

Even if he does not do all the work himself, a true handyman is a valuable member of the laboratory staff. He is able to spot the source of a problem and suggest repair methods. He is also able to come up with better ideas for handling laboratory modification or expansion.

14

A Case History

This is the step by step history of the planning and building of a medium size industrial laboratory. The progress and problems involved in this typical case history will serve to illustrate many of the more general or theoretical examples given in previous chapters.

The company was a private label manufacturer of home maintenance and personal care products. Its laboratory would be involved with new product development, evaluation of raw materials, testing of competitive products, and quality control. Laboratory personnel would also be responsible for chemical safety in the plant and for proper waste disposal.

SELECTING THE LOCATION

The company had taken over an empty industrial building. A section of the building close to the production area was chosen for the laboratory. Communication would be easy and interference from heat, noise, fumes, and vibration was judged negligible. This area had no utility connections, however, so another area closer to utilities had also been considered but was dismissed as being too far away from both office and production areas. Since utilities were required for use in the nearby production area, installation in the laboratory would not be too costly. The laboratory was to be a building within a building, with the "roof" creating a large mezzanine for storage under the much higher plant ceiling.

PRELIMINARY PLANNING

The planning and design were carried out by the engineering firm who also built the plant and who had on its staff a designer with long experience in laboratory planning. They worked in close cooperation with the laboratory operator.

The total floor area available was a little under 1500 square feet, part of which would be taken up by a first aid room and by a stairway and hallways. This would leave about 1250 square feet for the laboratory itself.

The laboratory operator's estimate of bench space needed showed that the space was adequate. Twice as much space as that needed for fixed equipment was requested, but it took time to convince management that this was in no way extravagant.

It was then decided that there should be a large main room and two smaller rooms, one for instruments and another for future microbiological work. The latter was hard to justify to management, so it was planned as a utility room for testing laundry products and working with dusty samples. It could be converted if the microbiological work came about. Meanwhile, the room was there, complete with utilities. In addition, there was to be an office, a closet area, and the first aid room mentioned above.

Many different arrangements were considered and rejected before a plan was developed that would be both practical and economically feasible.

Room dimensions finally settled upon were roughly as follows:

Main room	22x37 feet

Utility room	11x14 feet
Instrument room	10x22 feet
Office	9x11 feet
First aid room	8x11 feet

The rest of the area was made up of hallways, stairways, and a large size closet. Stairs to the mezzanine were designed to be wide enough for carrying large equipment up for storage. The main door to the laboratory opened into a small hall at the stairs, not directly into the plant, in order to avoid collisions with moving plant equipment.

The laboratory operator laid out work bench areas to suit jobs to be performed. Several changes were made in order to facilitate utility installations.

It was decided to give the large room a 9 foot ceiling, while ceiling height in the other rooms would be the standard 8 feet. Adequate space would be allowed above the ceilings to carry overhead utilities and air ducts.

DETAILED ROOM LAYOUT

Supplied with scale drawings of the proposed rooms, the laboratory operator began work on bench layouts in close cooperation with the designer. The designer made many practical suggestions, such as moving one of the sinks a short distance to simplify the plumbing. The large room was to have a work bench along most of one large wall, leaving space for the main door. Work benches would cover one of the short walls entirely and about 25% of the other long wall. The rest of this wall space would be used for doors to the instrument room, office, and the first aid room, and for a table for an analytical balance. The other short wall was left clear for a refrigerator and shelving for sample storage.

The large room would also have a center work bench. Utilities

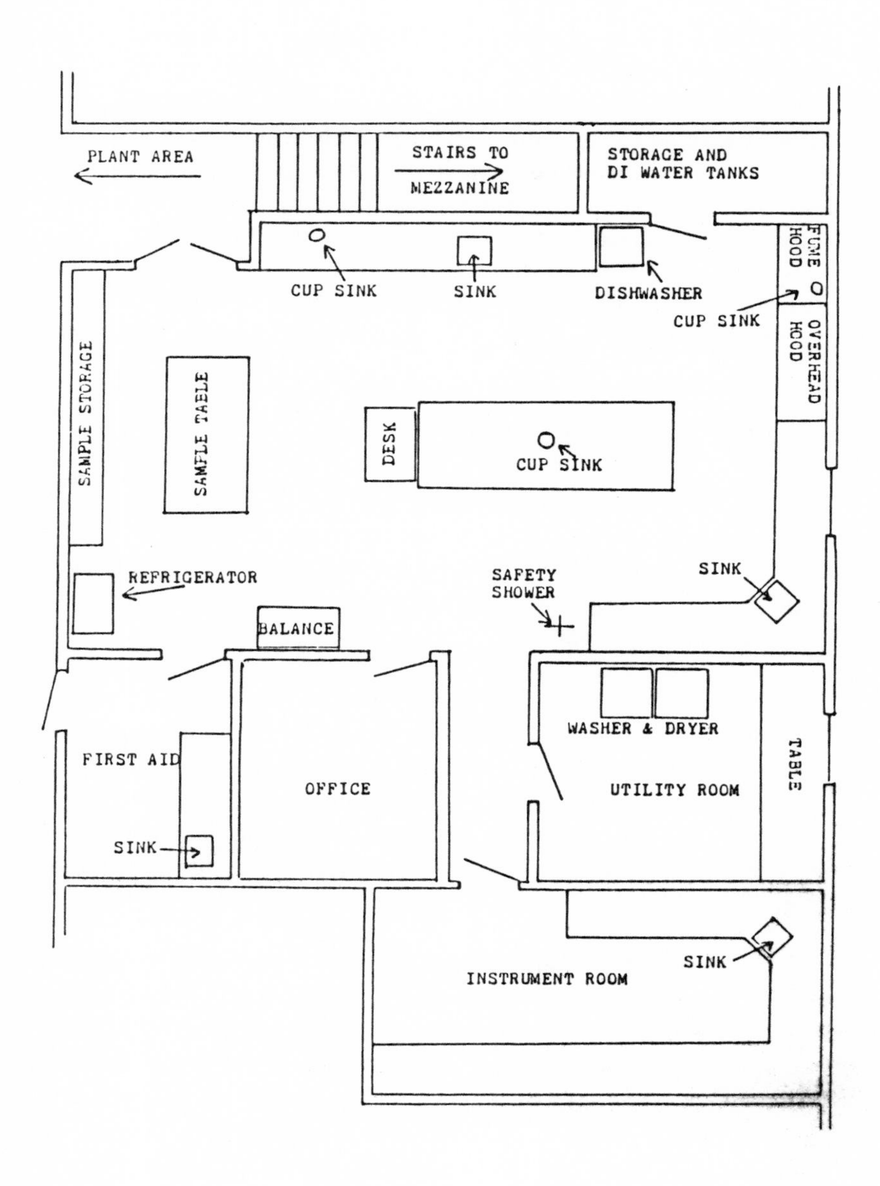

Figure 2: Laboratory floor plan as submitted to engineering firm.

would come to it from overhead, and the planned sewer would be directly underneath.

There was also to be room for a large table where incoming samples for analysis could be unpacked, sorted, and recorded. A suitable size was judged to be 4x8 feet. In addition, there should be room for a small desk at the end of the center table.

The instrument room was planned for work benches covering roughly 60% of the wall area. One wall was left free for bookcases.

UTILITY REQUIREMENTS

The laboratory needed the usual hot and cold water, electric power, gas, and sewer. In addition, there would be a need for DI water. Plant equipment was serviced with clean compressed air, which could be piped into the laboratory and be used with only minor further purification.

Except for the sewer, all utilities could be easily piped into the area. A water heater would have to be istalled, however, since the nearest one in the building was some distance away and already used to capacity. The concrete floor would have to be broken up to accommodate the sewer. A sewage mixing tank would be installed beneath the floor with access for sampling by a waste water inspector.

The building had an ample supply of electric power. It was just a matter of bringing the right amount to where it would be needed. The laboratory operator made careful estimates of what was needed and submitted his figures to the electrical designer, who added what was required for other items, such as lighting, and came up with a recommended number of circuits. The laboratory operator then added to this a separate circuit to be used exclusively for instruments using low but steady power, a matter the designer had not considered. In addition, the laboratory operator requested a breaker box that could accommodate several more circuits for possible future expansion. In the eyes of management, this was felt to be extravagant, but it was approved after some arguments. Most work areas would be supplied with power from wall-mounted strip-type outlets, but in some areas bench-mounted pedestals had to be used.

VENTILATION

The laboratory area had no ventilation, and the system used elsewhere in the building was unable to handle the extra load. As a result, a separate system had to be installed. At the laboratory operator's insistence, this was to be a system with no recirculation of air. There was considerable resistance from management due to the higher cost for both installation and operation. To make matters more difficult, the heating engineer had never seen a need for such a system on previous jobs. One argument finally settled the situation. It was pointed out that fragrance evaluations would often have to be performed as part of product evaluation. This would be difficult if much of the air were recirculated.

There were plans to build a small office for production personnel on part of the mezzanine, so the system had to be able to handle this, too.

The engineer recommended a heat pump for both heating and cooling mounted on the building's roof. To augment the heat pump in cold weather, electric heaters were installed in the ducts. The laboratory operator requested an on-off switch for the system with a pilot light mounted directly inside the laboratory door. Instead, a cumbersome timing device was mounted on the mezzanine, a "standard" procedure. It would have to be set for manual operation whenever somebody came in on a weekend. Nor was provision made for slow speed ventilation at night, as requested.

In retrospect, it can be said that the system had more than its share of problems. First of all, management refused a request for a maintenance contract, arguing that this system would work for a long time without maintenance. It did not. After about one year, the crucial reversing valve broke down. The unit was still under warranty, but this covered only parts. Labor charges were substantial. Two years later, the valve broke down again, but management still refused to consider a maintenance contract. Finally, one of the two fan motors ceased to operate, possibly because of lack of maintenance. This resulted in another big bill. On the positive side, it must be said that the thermostat system worked very well, although the installation of electric duct heaters may be questioned. They were probably not required in the mild California climate.

LIGHTING

A designer of lighting systems was called in. He recommended surface mounted fluorescent fixtures, which he considered more efficient than the recessed ones. The laboratory operator scanned all work areas with a light meter after the job was completed. The light was even and well within recommended levels. Furthermore, there were never any problems with the fixtures themselves, except for an occasional change of tubes.

LABORATORY BENCHES AND WORK TOPS

The laboratory operator had a preference for steel furniture, which the engineering firm shared. Rather than accept their recommendation for a manufacturer, however, the laboratory operator accumulated a stack of catalogs and studied them carefully. There was a substantial price difference between brands and their features varied. One maker seemed to offer the right combination between price and desired features. The sales representative arranged for the laboratory operator's visit to where this furniture had been in service for about two years. It was carefully inspected with slamming of doors and opening of drawers. Although the finish had at times been exposed to chemicals, it had held up well. Best of all, the work benches were available on short notice. After much deliberation, a cheerful two-color scheme of yellow cabinets and orange doors was selected. Installation by an independent contractor was included in the price.

In the main room where a large variety of work would be performed, bench tops with good resistance to chemicals and heat would be needed. In the instrument room, where strong chemicals would not be used, such tops would not be needed. The engineering firm, which had never built a laboratory where all the tops were not of the same type, was surprised at the laboratory operator's suggestion to use different materials for different work areas. Having just visited several quite sophisticated laboratories in Europe where ceramic tile was used, the laboratory operator was convinced that tile might be the most cost-effective option for the main room. Quotations were obtained from local firms. While much more expensive than plastic laminate, tile was far less expensive than monolithic composition tops. Furthermore, it was available from local suppliers on short notice. The engineering firm turned the idea down, however, and convinced management that the epoxy impregnated slabs were the only thing to use in the main laboratory room.

They did compromise by using laboratory grade plastic laminate in the instrument room, where chemical spills were rare. It proved to have a higher resistance than anticipated. The monolithic tops, as expected, took a great deal of punishment with no damage. The laboratory operator, though, would have settled happily for the far less expensive ceramic tile.

SINKS, FAUCETS, AND DI WATER

For the three sinks, the laboratory operator suggested standard enamel sinks. The more expensive epoxy sinks, however, were chosen by the engineering firm because of their high resistance to chemicals. The fact that the city had very strict regulations with regard to discharge of chemicals did not seem to alter their recommendation. The plumbing contractor presented several types of sink traps, from which the laboratory operator picked the one he felt was most easily serviced.

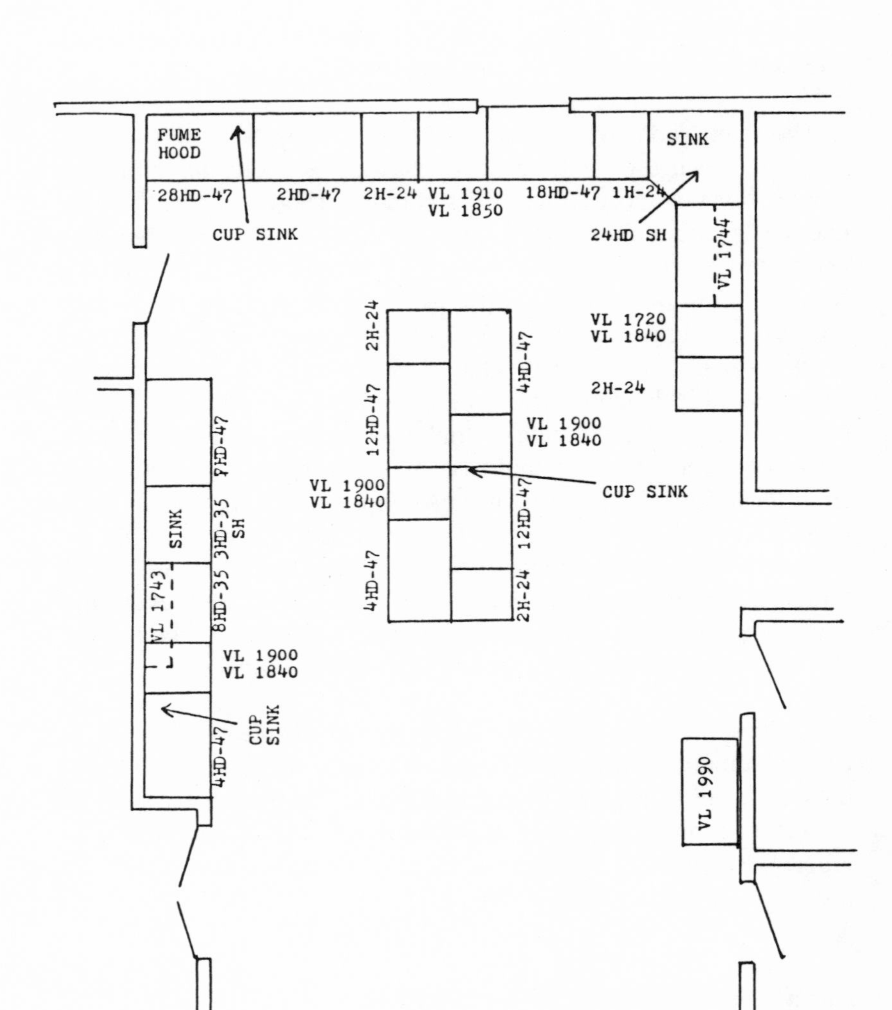

Figure 3: Plan for main laboratory room showing furniture modules with catalog numbers.

Faucets chosen by the engineering firm were equipped with anti-siphoning devices. The laboratory operator had specified their locations but had failed to check whether the local code required the expensive anti-siphoning type. The bench-mounted outlets installed for use with condensers and similar devices, however, were not of the anti-siphoning type, even though in this particular laboratory, the latter would be far more likely to draw water back into the system in case of pressure failure.

The engineering firm insisted on heavily plated metal valves for DI water. They had never used plastic, which the laboratory operator had successfully employed in a previous installation. Management agreed with the engineers. The water was distributed to all sinks by standard PVC pipe, which worked very well after thorough flushing.

Two DI units were employed in series and placed in the large closet next to the main laboratory room, where the indicator lights could be easily observed. This location was poor, since service personnel had to wheel replacement tanks through the laboratory, and mopping up was always required after exchanging tanks in the cramped quarters. Ideally, they should have been placed in the plant area with remote indicator lights mounted on the laboratory wall. At the time, such remote lights were not offered.

FLOORING, PAINT, AND CEILING TREATMENT

After some deliberation, sheet flooring was chosen rather than tile. Commercial grade vinyl was used in a color which harmonized with the laboratory furniture. The flooring was installed before any furniture arrived. While this called for more material, labor cost was much less. A sealer was applied to all seams. Although it stood up quite well, flooring with welded seams, not mentioned by either engineer or contractor, would no doubt have been more desirable.

The laboratory operator specified semi-gloss alkyd paint, despite the fact that both the engineering firm and the painting contractor claimed that a latex paint was what they always used for such applications. He also gave the painting contractor a narrow choice of high quality brands. While the job was underway, one of the painters remarked that it was nice to work with a good paint for a change.

Acoustical tile was specified for all ceilings, with the engineering firm selecting the appropriate type. Performance was good, the tiles seemed to have a low dirt pick-up, and dust deposits near the ventilation outlets were minimal.

FURNITURE INSTALLATION

The installer engaged by the furniture supplier did a truly professional job. It was finished sooner than expected. There was some fear that the monolithic countertop slabs, which were pre-fabricated at the factory, might have size discrepancies. They did not, due to very careful measurements taken when the order was placed. The plumbing contractor chosen by the engineering firm did excellent work. The electrical contractor was pleased with the way circuits had been planned, and the planner in turn was highly satisfied with his installation. At the laboratory operator's request, all information regarding installation and maintenance was saved and later filed. The contractors admitted that this was not their common practice, but they were all in favor of it.

EQUIPMENT AND SUPPLIES

Two large distributors were considered for providing equipment and supplies. It soon became apparent that neither could provide all the items needed, so it was decided that the order should be split.

It was an agonizing job to make up lists of everything needed, down to tweezers and filter paper, but the laboratory operator was aware that prices would be more favorable in such package deals. He also knew that management would be opposed to buying forgotten start-up supplies later on, usually at higher prices and with funds coming out of the operating budget. At the same time, he had been instructed to keep quantities of consumable materials to a reasonable level.

Eventually shopping lists were finished and sent out for quotations, which came as a pleasant surprise. Prices were well below those shown in the catalogs because of quantities ordered. Since the laboratory had not yet been finished, delivery dates were coordinated with estimated requirement dates. Manufacturers' representatives, from whom information on instruments had been requested, were quick to show up. This resulted in many free lunches, but not one representative resorted to what is usually called "hard sell". They were all very professional.

The large (4x8 feet) sample sorting table was custom made in the laboratory operator's carport. It was built to hold standard two-drawer file cabinets below. The top laboratory grade plastic laminate was procured from a local countertop fabricator. The file cabinets were spares from the office. A large office desk for the laboratory operator was obtained as part of the company's new office furniture. A small desk needed in the main laboratory room was purchased from a used office furniture dealer at a good price. The top was perfect, the drawers worked well, but the drab grey finish was scratched. A local paint store quickly matched the two colors of the laboratory furniture. There was a spray gun available at no charge. On a Saturday morning, the laboratory operator set up his paint shop in the plant area and went to work on the desk and the file cabinets. On Monday morning, the masking tape was removed and everything went back to the laboratory. It was well worth the modest investment in time and materials.

Since the company was getting new typewriters for the office, an older model electric machine was to be discarded. It was just right for the laboratory, but the laboratory operator insisted on having it serviced before taking it over.

Several home-type appliances were required, including refrigerator, dishwasher, washing machine, and dryer. A local appliance store provided them at a good price, much lower than if they had been purchased individually.

The company had declared surplus two secretary chairs because they did not match the decor of the new offices. They were in excellent condition and very comfortable. One was placed at the analytical balance, the other at the desk in the main room.

SAFETY CONSIDERATIONS

In case of an emergency, there were two ways to escape from the laboratory: through the main door or through the first aid room. The main drawback of this arrangement was that both doors opened into the same plant area. A better choice would have been to have installed a door from the instrument room to the office area.

Work areas were laid out in such a way that flammables would not be handled near either door. Upon arrival, cartons containing samples or supplies went where they would not interfere with traffic. Approved warning signs were posted where needed.

A combination safety shower and eye wash station was installed in the laboratory. Although the engineering company recommended a very expensive chrome-plated unit, a much lower priced one which used plain galvanized pipe was chosen. Its performance was the same. The pipe was later primed and painted to match the laboratory furniture, another Saturday job for the laboratory operator.

The building's sprinkler system, already in place, was extended to the laboratory area by a properly certified contractor.

The fire department was consulted with regard to the proper number and placement of fire extinguishers. They were most happy to cooperate and had some good suggestions. Management agreed to a service contract for the extinguishers.

CONCLUSION

As with any other project, there were things that could have been done differently and better a second time, but overall it must be said that, in spite of occasional differences of opinion, this laboratory installation worked out well. It was the cooperative effort of many people, each one doing his or her best with an eye for the future. No dramatic changes had to be made as work began or when new assignments came in. The laboratory became a place where creative and productive work could be performed with minimum interference.

Index